Undoing Whiteness in Disability Studies

Sana Rizvi

Undoing Whiteness in Disability Studies

The Special Education System and British
South Asian Mothers

Sana Rizvi
Graduate School of Education
University of Exeter
Exeter, UK

ISBN 978-3-030-79572-6   ISBN 978-3-030-79573-3   (eBook)
https://doi.org/10.1007/978-3-030-79573-3

Cover illustration: © Alex Linch_shutterstock.com

This Palgrave Macmillan imprint is published by the registered company Springer Nature Switzerland AG.
The registered company address is: Gewerbestrasse 11, 6330 Cham, Switzerland

This book is dedicated to
Alina
Kiran
Maham
Maria
Parveen
Saira
Shehnaz
Tahira
And
Ali and Reza

FOREWORD

Intersectionality offers a window into thinking about the significance of ideas and social action in fostering social change. Although intersectionality has been consistently aligned with visionary ideals such as freedom, social justice, equality, democracy, and human rights, neither change itself not intersectionality's connections to such change is preordained. The only thing that is truly certain about human existence is that it will change, but not necessarily in the evolutionary, linear fashion of Western notions of progress. [...] During times of such visible and contentious change, it's reasonable to question the worth of intellectual work, especially when everyday social problems seem so pressing. Yet because we've been here before, we cannot search for certainty, but rather for critical analytical tools what will enable us to grapple with the ever-changing contours and durable effects of social problems. (Collins, 2019, p. 286/287)

And as for God? I know that for a start, anger at God is a legitimate aspect of spiritual life. To be angry at God is just as much a part of prayer as is love or devotion, awe or repentance. In all those states of mind, feeling and soul, we are confirming that moral laws bind the universe as a whole, just as they bind us as moral agents within it. That's why I think I'm entitled to rage at God when I witness Esther's pain, her limitations and disappointments. When I think that "normal" love relationships and opportunities will simply never be hers. As she herself said, when I told her that because of her special needs she could not return to her beloved day camp as a counsellor: "Now's the time to scream at God." (Gottlieb, 2002, p. 11)

Perhaps the most fruitful distinction with which the sociological imagination works is between 'the personal troubles of milieu' and 'the public issues of social structure'. This distinction is an essential tool of the sociological imagination and a feature of all classic work in social science. (Mills, 1959, p. 8)

I begin this foreword with three quotes perhaps because they encapsulate much of what is going on in the ground-breaking book ahead: theoretical positioning via the lens of intersectionality and Black Feminism, how culture and religion impact upon the personal lives of disabled families, as well as reflecting upon the broader public sphere for all concerned. As Rizvi says in Chap. 7, "*It is a challenging task to change how local authorities and official institutions—that are predominantly White and middle class—view and work with minoritised families, unless we push for radical change*".

Critically, Rizvi carries out this work via an intersectional lens influenced by Patricia Hill Collins and Sirma Bilge (2017). Rizvi is delving into the deeply emotive narratives of stigmatised maternal and disabled, everyday life, whilst contextualising the micro and macro racialised cultural and religious landscape for British South Asian Mothers of disabled children. Hence in discussing disability, mothering, 'special educational needs' (SEN), culture and religion, it would be negligent to attempt to understand any social tradition or norm, without considering its biography, history and social structure. This is precisely what Rizvi does. Not least because becoming a mother to a disabled child radically changes the expected horizon of what being a mother involves, her public performance and her private internal *and* external dialogues—these are evident in the book ahead.

As expectations of cultural norms, whether it is celebrating a birth, maternal bonding, returning to work, a child's healthy body and mind, socially and sexually appropriate behaviour, academic ability, access to community and religious activities and schooling, means that some of the above expectations are shattered and a normative life course changes forever. Notably, the trauma for families with disabled children is already too much to bear, as a recent report based on family research (Challenging Behaviour Foundation (CBF), 2020) highlights. As a reader of Rizvi's work ahead, and scrolling through the pages of the CBF report, it is evident, that what is already incredibly traumatic for families, for example, children being left in their own soiled underwear, or more generally a lack of specified and identifiable support, is further compounded by racialization and the cultural and religious landscape that families in this book navigate. Furthermore, in the CBF study, it is manifest 'Covid-19 has highlighted the inequalities in our society, even more. The restrictions imposed create settings where the risk factors identified by families as increasing trauma are more likely to be prevalent' (CBF, 2020).

This book is an in-depth story telling of eight British South Asian mothers who have at least one child who is disabled. It is based on a rich qualitative study that in-detail and over three interviews per participant, traces the lives of these mothers. Rizvi highlights the micro-politics, the private concerns and everyday desires of these mothers and their disabled children, yet she shines a light on the broader socio-political terrain of significant flaws in health, education and social care policies and practices for British South Asian families. It is clear that there are assumptions about knowledge in terms of the local SEN environment, and a lack of parental voice. These assumptions and this neglect leads to a deep vacuum in any meaningful negotiation of what is necessary for support for disabled children and their families. Yet it is evident in policy documentation that parents are to be part of discussions about their children and yet are continuously missing, left out and ignored. Moreover, in mothering a disabled child there are significant costs. These costs straddle the emotional and practical reserves of these mothers. However, none of these 'costs' are easily defined due to the messy nature of the emotional work, love labour and care work involved. In addition, whilst the act of becoming a mother might be considered personal—a private occurrence—it is increasingly a site of public surveillance more often via the health, education, and social care infrastructure.

As it is, classed narratives around missing voices, maternal competence, and the deficit family has been going on for some decades in terms of mothering and disability as highlighted by Rizvi. Yet as Rizvi begins her telling of this story, she pointedly highlights the sheer additional cultural and racialised assumptions that British South Asian mothers with disabled children are traversing. For example, in Chap. 1, she describes her first meeting with a school home-liaison officer, Alice, and it is worth repeating some of this here. Alice told Rizvi that the school was outstanding, and yet had some problems around parental engagement with families from certain communities. Rizvi asked for some clarification on this, and Alice said about parents from Bangladeshi backgrounds,

... it appears they are just indifferent. We've tried coffee mornings, parent groups ... I mean they just don't attend any of that. I personally think it's because their women, they are shut up in their houses aren't they? They're oppressed, they need to get out of their houses! They're living with large families, with their in-laws and all their relatives, where's the privacy? And you know they're all related ... [in a lowered tone]. I'd get into trouble for saying this, but we all know why they have babies with disabilities. I can't understand how you can marry your cousin and live with his entire family ... where's the romance in that?

Rizvi goes on to say she thought that Alice was joking with her, and said, '*I'm married to my second cousin, and I live with my in-laws and his siblings*'. This opening of the book biographically encapsulates in an emotive way how damaging racialised cultural and religious stories play out. Indeed, the concept of racialization extends back to the nineteenth century, but in more recent history became known as the 'race relations' paradigm in the UK (Solomos, 2019). Whilst different ideas around racialization have evolved, especially by Frantz Fanon (1961) where it is synonymous with dehumanisation, others have described the contested nature of racial categories and race-thinking processes, especially in European colonial encounters that pursued the legitimation of white supremacy. It is clear therefore, that the British Asian mothers with disabled children are caught up in a number of damaging discourses that are racialised, disabled, stigmatised, challenging, disruptive, non-compliant and deficit—indeed dehumanising. By studying the personal and private lives of these mothers in the context of their cultural and religious lives, questions and answers about the broader social picture can be asked and addressed, as what we learn in the following work is how these British South Asian mothers experience normative difference, and of the sheer practical and emotional work involved.

Notably, this book lands at a time when more Black, Asian and minority ethnic groups are harder hit by the global COVID-19 pandemic. For example, their mortality is higher than the average population due to several factors including socioeconomic inequalities, over representation in public facing employment and communication barriers. It is also at a time when those in the public eye, such as Radio DJ Jo Whiley, shines a light on the over representation of deaths of learning-disabled people from coronavirus and how they ought to have COVID-19 vaccine priority. It is clear that disability, special education and health cannot be disentangled from the broader socio-political landscape of cultural and religious identities and narratives. Furthermore, that personal narratives are heard and acted upon, rather than relegated to parochial and culturally specific satellites.

This book is so much more than the title could possibly reveal, as by providing a window into an overlooked and particular group of mothers, Rizvi evidences a nuanced perspective on mothering disabled children with an emphasis on schooling, education, cultural diversity, Muslim identity and religion. Significantly, she highlights the complex lives that these mothers live when, in addition to the 'ordinary' disabling challenges, we hear about how religion, consanguinity, culture and other difficult differences play out. Complicated and sensitive issues are raised, for example, around participation in mainstream education and religious activity, as well as sexuality and behaviour, and caring practices. In addition, the socio-political terrain is mapped as the reader gets a sense of bureaucratic (and potentially useful or damaging) processes such as Statementing and support, as well as how inclusion might work (or not) differently in the context of this group of mothers. As identified by Rizvi, the additional caring work due to the cultural and religious tensions causes further stress and anxiety. The religious spaces, however, can offer support not systemic violence, as she sees a cultural gap where mosques and religious centres could enable community support. It is therefore suggested that religion ought to be considered as a beneficial resource, not something that hinders rights and freedom. Returning then to intersectionality and its practical application, it is proposed as an analytical tool because the reduction of South Asian disabled children and their families to their culture and language is too big an issue to brush under the racialised carpet.

Canterbury, UK   Chrissie Rogers

REFERENCES

Challenging Behaviour Foundation. (2020). *Broken: The psychological trauma suffered by family carers of children and adults with a learning disability and/ or autism and the support required.* Retrieved February 8, 2021, from https://www.challengingbehaviour.org.uk/learning-disability-assets/brokencbffinal-reportstrand1jan21.pdf

Collins, P. H. (2019). *Intersectionality as critical social theory.* Duke University Press.

Collins, P. H., & Bilge, S. (2017). *Intersectionality.* Cambridge: Polity Press.

xii FOREWORD

Fanon, F. (2001 [1961]). *The wretched earth*. Penguin Classics.
Gottlieb, R. S. (2002). The tasks of embodied love: moral problems in caring for children with disabilities. *Hypatia, 17*(3), 225–236.
Mills, C. W. (1959). *The sociological imagination*. Oxford University Press.
Solomos, J. (2019). After Michael Banton: Some reflections of his contributions to the study of race. *Pattern of Prejudice, 53*(4), 321–336.

ACKNOWLEDGEMENTS

This book wouldn't be possible without my partner, Reza Rizvi. Thank you for your thoughtful insights, feedback, staying up late with me to keep me going, driving me to interviews and doing way more than your fair share of the household chores just so I could get time away for writing this book. It's not a miracle that I wrote most of this book during a pandemic; I have you to thank for constantly believing in me.

Thank you also to the late Dr Penny Lacey—I'm sure there are many people whose life you made so much better, but I'll forever stay indebted to you for your valuable advice, for writing endless references for me even though you were ill or you had a pending operation—and for pushing me to give PhD a serious thought. Thank you for modelling how to be a gentle academic.

I'm also grateful for the mentoring and the scholarship of the following people—Dr Neil Hall, Prof. Saba Fatima, Prof. Gemma Moss, Prof. Sarah Younie, Helen Knowler and Prof. Anthony Feiler—thank you all for listening to my grumbles and for always looking out for me. Academia needs more mentors like you all.

To my sisters—thank you all for your solidarity. Writing with multiple deadlines can be tough and I deeply value all my conversations, guidance, quick chats, various writing groups, late night pep talks and the wonderful scholarship of the following people—Rajvir Gill, Talitha Bird, Dr Gisela Oliveira, Angie Sibley-White, Dr Alison Black, Dr Sarah Cole, Dr Sharon Morgan, Dr *Hannah Anglin-Jaffe,* Zahra Bei, Dr Gayatri Sethi, Dr Altheria Caldera, Dr Freyca Calderon-Berumen, and Dr Betsabeth Lugo.

I'm deeply indebted to my sisters in the women of colour binder academic group and the Network of Sisters in Academia (NeSA). I've benefitted from their wisdom and support each time I had a doubt, a question, or needed emotional support. You all know who you are—I thank you all from the bottom of my heart.

My thinking and my work draw inspiration from engaging with the following amazing people—Prof. Sue Timmis, Prof. Sheila Trahar, Prof. Jo Rose, Dr Wan Ching Yee, Prof. Chrissie Rogers, Prof. Dan Goodley, Prof. Melanie Nind, Prof. Richard Hall, Prof. Alison Reiheld, Prof. Fatima Pirbhai-Illich, Dr Suma Din, Dr Hadiza Kere Abdulrahman, Dr Frances Ryan, Malcolm Richards, Riadh Ghemmour, Dr Jonathan Doney, and Dr Gabriela Meier. I thank each of you for your solidarity and the incredible work that you do.

I truly believe in the power of prayer—I have always felt supported knowing my family has prayed and rooted for me. My eternal gratitude and love go out to Reza Rizvi, Prof. Saba Fatima, Sultan Murad, Nasim Murad, Dr Hina Rizvi, Zia Rizvi, Dr Huzaifa Quaizar, Ali Azam, Sonia Shaikh, Ibrahim Murad, Dr Fatima Durre, Roohi Rizvi and the late Masood ul Hasan Rizvi. And special (and additional) thanks go to Dr Hina Rizvi, Zia Rizvi and Reza Rizvi for looking after my Ali *juju* so I could write.

Last but not least, my sincere appreciation to Charanya Manoharan, Rebecca Wyde and Eleanor Christie at Palgrave Macmillan—thank you for supporting me every step of the way.

CONTENTS

ABBREVIATIONS

ADHD	Attention Deficit Hyperactivity Disorder
ASD	Autism Spectrum Disorder
BAME	Black and Minority Ethnic
BIPoC	Black, Indigenous, People of Colour
BCODP	British Council of Disabled People
BESD	Behavioural, Emotional and Social Difficulties
CAB	Citizens Advice Bureau
CAMHS	Child and Adolescent Mental Health Services
DLA	Disability Living Allowance
DPM	Disabled People's Movement
EHC	Education, Health and Care
ESOL	English to Speakers of Other Languages courses
FBV	Fundamental British Values
FSM	Free School Meals
GDPR	General Data Protection Regulation
IEP	Individual Education Plans
LA(s)	Local Authority (Authorities)
LGBTQ+	Lesbian, Gay, Bisexual, Trans, Queer/Questioning (plus sign reflects a desire to be inclusive)
MCB	Muslim Council Britain
OT	Occupational Therapist
PMLD	Profound Multiple Learning Difficulties
PEG	Percutaneous Endoscopic Gastrostomy
PTA	Parent Teacher Association
SEMH	Social, Emotional and Mental Health
SEND	Special Educational Needs and/or Disabilities

SENDisT	Special Education Needs and Disability Tribunals
SLD	Severe Learning Difficulties
SLT	Speech and Language Therapist
SSLD	Specific Speech and Language Difficulties
VBA	Verbal Behavioural Analysis
UPIAS	Union of the Physically Impaired Against Segregation

PART I

CHAPTER 1

The Harms of Being Spoken For

What Is the Dominant Perspective?

It is a strange feeling to start this book with an incident that happened nearly a decade ago. But the incident captures slivers of the types of interactions that I have lived through in so many different professional settings, academic circles and read within literature on disability. It was spring in 2011 and I was nervous and excited in equal measure to be finally meeting with the special school, located in Birmingham, which had agreed to participate in my research. Up to now four schools had politely refused, citing their busy schedules at that time of year. I had never been inside a special school in Britain before because my entire educational upbringing had been in Pakistan. Understandably, I was anxious to make a good professional impression. This was to be my first research experience with participants from a British South Asian background. I was scheduled to meet Alice, the school's home-liaison officer that day.

The following is a brief extract of the conversation I had with Alice which I noted down in my research diary.

Me: *In what area do you think my research could benefit your school the most?*

Alice: *Erm, well as you know we're an 'Outstanding' school so we're already quite good with certain aspects of schooling. I guess there is a problem*

© The Author(s), under exclusive license to Springer Nature Switzerland AG 2021
S. Rizvi, *Undoing Whiteness in Disability Studies*,
https://doi.org/10.1007/978-3-030-79573-3_1

as far as parental engagement is concerned with families from certain communities.

Me: *What do you mean?*

Alice: *It's parents from Bangladeshi backgrounds more than any other community, we need to engage them more, but it appears they're just indifferent. We've tried coffee mornings, parent groups … I mean they just don't attend any of that. I personally think it's because their women, they are shut up in their houses aren't they? They're oppressed, they need to get out of their houses! They're living with large families, with their in-laws and all their relatives, where's the privacy? And you know they're all related …* [in a lowered tone] *I'd get into trouble for saying this, but we all know why they have babies with disabilities. I can't understand how you can marry your cousin and live with his entire family … where's the romance in that?*

I felt amused, thinking that Alice was joking.

Me: *Actually …* [I pause before fulfilling her stereotype about South Asians] *I'm married to my second cousin, and I live with my in-laws and his siblings* [I laugh nervously].

Alice: *Oh my God!*

We were both laughing now.

Alice: *So how do you get privacy you know to …?*

Me: *It's not as bad as you think, it's like we live together but we're not in each other's faces … so are you interested in knowing why Bangladeshi parents aren't coming into school?*

As I walked back to my car, I felt satisfied that we had 'located' an area of research, even if our conversation had been uncomfortable. It seemed to me that we had both gotten something out of this interaction; she had reaffirmed her problematic beliefs that all South Asian Muslims have cousin marriages and live together with their in-laws in crowded homes, whilst I had gotten access to my participants which I would not have otherwise. There was so much that bothered me about this conversation. I reflected on why I felt embarrassed about my marriage—what was wrong with South Asians marrying in their kin? What is wrong with living in joint family system with one's in-laws? The discomfort that I felt during this meeting has since then become a good reminder of how as scholars of colour trained in Western ways of researching minoritised communities, there is a risk of being *"implicated in* [our] *own oppression and marginalisation through a form of false consciousness"* (Atwood & López, 2014, p. 1137). There is a danger, as Atwood and López (2014) warn us, of not questioning an imposed narrative. As an early career researcher at the time, I did not have the presence of mind to articulate the 'unsaid' but necessary point that this situation demanded. That is, for an experienced home-school liaison officer there was a lack of understanding and insight of how ethnically minoritised communities do not have the privilege of generational wealth to own housing, nor the understanding of the various mechanisms of institutional racism that hinder ethnically minoritised communities from becoming home-owners; or that perhaps living within a joint family system allows young couples to save up and own housing themselves in the future. There was a lazy assumption that large families could only be seen as 'crowded' which invaded the privacy and quality of one's personal life. In a single encounter, the home-liaison officer had managed to cluster labels such as 'oppressed', 'large families', cousin marriages and lack of romance to define the local British Bangladeshi community. It is staggering how swiftly and recklessly the multi-layered concerns of this community were boxed into her preconceived notions.

It is difficult to pinpoint exactly when I identified that my current research interest would be minoritised maternal engagement and experiences, because it was a gradual realisation. This one incident reflects deeper issues within various discourses surrounding the different ethnic minoritised groups—in particular the ethnic minoritised mothers, and the manner in which these discourses seem to have trickled into research on home-school relationships and parental engagement within the field of special education and disability studies. In 2011, I wrote the following in my field notes:

Did Alice already have preconceived notions about why Bangladeshi mothers were not coming into schools? The fact the school had organised coffee mornings and parent groups, was this sufficient reason for Alice to believe that they had fulfilled the school's responsibility towards Bangladeshi parents? Did Alice actually believe that Bangladeshi parents could be blamed for having disabled children, simply because some South Asian Muslims had cousin marriages? With my Western attire and educational background, did Alice assume that I was an exception, an ally? Do staff members discuss certain ethnic groups in certain ways behind the closed door of their staffrooms?

In hindsight, I reflected upon my awkward defensiveness and on how this had seeped through into my writing and my way of thinking.

WHAT DO WE KNOW ABOUT THE MOTHERING OF DISABLED CHILDREN?

In recent decades, a few studies have explored the parenting of disabled children from a maternal perspective (Landsman, 2005; Ryan & Runswick-Cole, 2009; Rogers, 2011; Knight, 2013). Notably, how mothers are represented by researchers, whether it is through a gender-evasive lens[1] or a 'super mum' perspective, it is arguably different from how the mothers would portray themselves. Nonetheless, these studies have been instrumental in exposing the patriarchal pathologising nature of home-school relationships. Existing studies on home-school relationships (Fylling & Sandvin, 1999; Stoner & Angell, 2006; McCloskey, 2010), have highlighted how parents may be problematized and categorised as inadequate or noncompliant if professionals perceive them as failing to perform their roles within the parent-professional relationship. Since parental involvement discourse is overwhelmingly framed through a professional lens, it is unsurprising that a parent who does not represent the 'ideal' maybe labelled an 'angry-knowledgeable parent', 'submissive parent', 'fighting parent', 'uncaring parent', 'angry ill-informed parent', or 'special needs parent' (Gascoigne, 1995). Although these parent behavioural categories may have been developed to help professionals to look beyond their first impressions of parents, nonetheless, they risk scrutinising parental practices over and above other factors that may be affecting the parent-professional relationship. These categories also highlight who holds the power in imposing labels on others, and indeed how the 'professional' label affords a blanket protection to be perceived as objective and impartial. Parental involvement discourse is also overwhelmingly patriarchal, invisiblising mothers who are frequently at the forefront of these often-imposed professional relationships which might be oppressive to them and their families. It is, therefore, important to acknowledge how literature on

---

[1] The gender-evasive lens refers to how research on parenting may not recognise that most of the parenting and caring is disproportionately carried out by mothers.

parenting disabled children from a maternal perspective is pertinent to understanding how mothers view their own roles within professional partnerships, and whether they find these partnerships meaningful to their day-to-day experiences.

Within mothering literature, mothers of disabled children are also scrutinised with regard to how they conceptualise their child's disability and how they identify with publicly available discourses and models of disability. Currently, there are two prominent standpoints to understanding how disabled families engaged with disability and how they are supported by professionals: the medical and social models of disability. The medical model of disability, which is utilised by the World Health Organisation, conceptualises disability as a consequence of an individual's physiological and biological condition (Johnston, 1996). Landsman (2005) suggests that the medical model imposes strict binaries between what is normal and abnormal, and problematizes those individuals who challenge and question this imposed normative. Landsman's (2005) USA-based qualitative study included 60 mothers who had a child or were expecting a child with disability. She explored how mothers made sense of their child's disability, and how this conflicted with existing models of disability. She found that mothers who rejected the initial diagnosis of their child's disability did not reject medical professional expertise, but rather the 'disabled' label. Mothers noted that child assessments lacked contextualisation and inaccurately represented their child's behaviour and capabilities; these mothers were told by doctors to *"face reality"* (Landsman, 2005, p. 128) and accept their child's 'disability'. This illustrates how parents can become subject to a professional gaze, being perceived as being in denial according to the Grieving Model (Kubler-Ross, 1969) if they resist their child's diagnosis or their response falls outside of expected norms. In my earlier research (Rizvi, 2015), one mother recalled how doctors were surprised by her joyous reaction after they disclosed her child's medical condition. She was adamant that she was simply overjoyed at the birth of her first child and, to her, disability was not the primary marker; however, doctors strongly advised her to grieve if she needed to. This raises unsettling questions about how some medical professionals may consider the birth of a disabled child as parents being deprived of having a perfect child; moreover, mothers may be forced to identify with emotions that professionals have classified as a normative response.

Sousa (2011) suggests that the availability of prenatal testing, pregnancy specific dietary requirements and parenting classes has led expectant mothers who adhere to medical expertise to perceive that disability is 'preventable'. Moreover, mothers are aware that they are often blamed for their child's disability, which I also found in my research (Rizvi & Limbrick, 2015). As I have already discussed at the start of this chapter, some schools professionals do view South Asian marital practices between cousins as directly responsible for their child's disability, which affects the lens through which they engage with parents (Rizvi & Limbrick, 2015).

To some extent, the problematic imposition of a medicalised lens on how to experience the birth, caregiving and mothering of a disabled child has been challenged by the social model of disability (Campbell & Oliver, 1996). The social model proposes that it is the presence of external societal and structural barriers—which may take the form of ideological barriers—that prevent disabled people from fully participating within society. Maternal actions in Landsman's (2005) study reflected the messiness of engaging with different models of disability; for instance, they acknowledged that medical professionals would deem them good mothers for seeking 'cures' or interventions, but these actions would be viewed negatively by disability advocates as an attempt to change their child rather than transforming the wider system. Landsman found that mothers in her study recognised that it was wider society that was structurally disabling and felt protective of their child against disabling attitudes, wanting to present them as 'normal' as possible. Landsman's study has important implications; firstly, by seeking interventions, mothers may not necessarily be rejecting the social model but rather equipping their child to evade oppressive societal attitudes towards 'difference'. Secondly, utilising interventions to improve their child's life chances should not be viewed in contradiction to the values of dignity, inclusion and independence espoused by disability advocates. Ryan and Runswick-Cole (2008) suggest that rather than regarding mothering as being shaped by the medical model of disability, one should recognise that mothers have no other choice but to interact with the medical model in order to support their disabled children within rigid medicalised institutional practices. Mothers seek labels as a tool to not only secure better provisions for their children, but also to resist 'mother blaming' from the community and to smooth over socially difficult situations.

Scrutinising mothering through different models of disability can marginalise maternal experiences because their loyalties are pushed into either a medicalised or into a social 'camp', without acknowledging the lack of choice and agency that they have been permitted. Fisher and Goodley's

(2007) UK-based study with families of disabled children highlights how mothers have challenged such imposed choices and narratives. They found that whilst some mothers ascribed to a medical framework, they did so to eliminate uncertainty about their child's prognosis; mothers of children with rare genetic conditions utilised the medical model to adjust to their 'new normal'. Many mothers resisted the linear medicalised narrative, challenging normative values which they found oppressive and through this process realised their agency. Whilst the mothers in Fisher and Goodley's (2007) study ascribed to the social model of disability, they also supported and navigated their children's disability in a nonlinear manner. Rather than fitting into existing frameworks, mothers found different individual mechanisms that worked for them. Some mothers understood their everyday experiences through religion, and some mothers lived life in the present rather than worry unduly about their child's uncertain future. Fisher and Goodley (2007) included three mothers from a British Pakistani background as participants in their study and looked at religion as a way by which some mothers support their disabled child. Their work was instrumental in initiating this very important conversation, opening up an entry-point for delving deeper into how mothering occurs at various intersections such as religion, ethnicity, disability and other social categories.

What Do We Know About the Nature of Partnerships That Are Extended to Mothers?

The research by Rogers (2011)[2] is a seminal study in that it reflects the dynamics and power imbalances within mother-professional partnerships.[3] Rogers (2011) explored maternal experiences of engaging in professional partnerships, and how by entering into the public world of the special education system mothers were subject to the "*welfare gaze*" (p. 574). Rogers (2011) suggests that when mothers of disabled children seek welfare support, they expose themselves and their children to a system which defines them as needy and dependant which invariably affects their relationships with professionals, resulting in mothers being 'talked at' rather

[2] Rogers (2011) broadly defined 'mothering' as caring for a child with SEND which can be done by both mothers and fathers.
[3] The term 'mother-professional partnership' acknowledges that in most instances, it is mothers who interact and establish relationships with professionals.

than a mutual dialogue. The mothers in her sample suggested feelings of entrapment; they wanted to do everything possible to the best of their mothering abilities for their child, without feeling a lack of control or being overwhelmed by professional interference. Mothers from different socioeconomic backgrounds reported similar warlike experiences with professionals, feeling forced to adopt a 'fight' and 'battle' posture when they did not receive the expected level of services or support from professionals (Rogers, 2011). This experience of fighting and battling is intricately tied to the welfare gaze; for instance, one working class White mother reported that if she did not do her 'homework' on her caring role, 'they' would just side-line her child as nobody really cared. Mothers struggled to balance their exhaustion and frustration, fighting against their impulse to put their child into formal care, but rather to continue to fight publicly for everything possible for their child like 'a good mother' should. From a seemingly sociological and feminist lens, Rogers (2011) suggests that mothers may not always appear to be "*professional parents*" (p. 565), nor always willing to conform to education and health professional expertise.

This lack of 'professional' behaviour can often lead to mothers being blamed for their child's disability. This was reflected by the experiences of a middle-class White mother in Rogers' study, who suggested that her whole family was called into a family consultation meeting at school over her disabled son's difficult behaviour, and where she endured inappropriate questions which seemed to point all the blame for her child's behaviour at his home environment. The mother acknowledged that the questions may have been deemed appropriate from a professional viewpoint, and that she and her family may have appeared uncooperative because they was uncomfortable at being unceremoniously surveyed and ostensibly blamed for all the child's behavioural difficulties (Rogers, 2011). The example also reveals the double consciousness that mothers engage in knowing how their public behaviour is perceived during these meetings, and feeling defensive when their mothering skills are being scrutinised and questioned. Families who did not 'cooperate' were categorised by professionals as dysfunctional and needing "*intensive family support*" (p. 576). Rogers (2011) also suggests that a school's demeanour towards a mother may be influenced by her child's disability type. Children with social, emotional and mental health (SEMH) needs are less visible compared to other disability types, inspiring less empathy from professionals and the general public and being more often disciplined. Consequently, schools which view children as 'problematic' or 'difficult' can in turn also discipline the mothers rather than provide them support.

Rogers' study (2011) is significant because it challenges the notion that motherhood and mothering a disabled child is an entirely natural act for all mothers. It also revealed that any maternal apprehensions or feelings of despair, which fall outside the cultural norms and expectations representing good mothering, immediately invited professional criticism as well as normalisation techniques from various institutional settings (Rogers, 2011). It presented mothers who were at times uncertain about their mothering skills, and were unsure how the 'partnerships' with various professionals constituted true partnering, making them more insecure still about themselves and their mothering skills.

WHO IS THE FOCUS IN THIS BOOK?

I feel that I need to explicitly state how to approach reading this book. This book is not a window into understanding children with disability. It is not that this topic is not an important subject to study—on the contrary, minoritised disabled children and young people's perspectives are central to understanding how they experience inclusion and exclusion within education and wider society (Hussain et al., 2002; Annamma, 2018). Rather, my interest lies in utilising the intersectional lens to contextualise and focus on the mothering of disabled children from the British South Asian Muslim community. There is a myth perpetuated by media, academic and government discourse in this country about the 'culture of disengagement' within this community; British South Asian Muslim mothers are either perceived as passive stakeholders who do not engage with professionals, or as a threat to the integration of their children into mainstream society. This book is, therefore, about the perspectives of British Pakistani mothers on mothering their disabled children and their experiences of the special education system. By special education system, I am referring to all the micro-level informal and formal settings, provisions, processes and interactions with different stakeholders and professionals, as well as macro-level Special Educational Needs (SEN) policies and other contexts that shape their child's experiences of disability. There is no doubt that, at times, the maternal narratives in my study will be perceived as being deeply embedded within patriarchy and ableism, and it is important to acknowledge that mothering is a complex and an often-contradictory task.

Scuro (2017), a professor of philosophy and a mother of a disabled child, asks whether parents can ever be allies to their disabled children. This is an important question to address for any book that focuses on the parents of disabled children, especially if one considers how many professionally developed parenting models internalise ableism. Scuro (2017) insists that the purpose of asking this question is not to parent-blame nor is it to demonstrate that some ways of parenting are better than others. Rather, it is to question how parental expectations, roles and responsibilities are often shaped by an ableist discourse. As mothers of disabled children, Ryan and Runswick-Cole (2008) suggest that the role of mothers as either 'allies' or 'oppressors' has been deeply debated within disability and feminist studies. From a disability studies viewpoint, there are several reasons why a focus on the role of mothers and mothering of their disabled child may be perceived as oppressive to the child. For instance, focussing on maternal perceptions and experiences may be construed as shifting the focus away from the needs of the child, or possibly even perceiving that mothering a disabled child is burdensome. This view risks only focusing on the negative aspects of raising a disabled child and does not acknowledge them as persons in their own right. It has strong implications from a human rights perspective, because if disabled people are characterised as being wholly dependent on caregiving assistance from close family members for their personal needs, it makes them more vulnerable to the caregiver's whims and goodwill and inevitably creates an imbalanced power relationship (Morris, 2001). Power dynamics have also been highlighted within research exploring religious explanations for disability amongst parents, which considers caring for disabled children as a religious calling and almost inevitably risks viewing their children through a charitable and altruistic lens (Ryan & Runswick-Cole, 2008). In addition, the ability of nondisabled mothers to raise disabled children has been questioned, particularly whether these mothers try to impose normalcy upon their disabled child (Middleton, 1999). The less severe a child's disability is, the greater is the maternal effort to 'normalise' the child. This examination is more evident in research with mothers who have children with hearing impairments, specifically with respect to their decision of whether or not to pursue Cochlear implants for their children (Hyde et al., 2010).

Coming back to Scuro's (2017) question—can parents ever be allies to their disabled children?—the answer is not simple and needs to be contextualised. For instance, we need to consider how the supposedly 'universal' conceptualisation of ableism may itself be Western-centric, requiring a rethink to take into account how minority parents can become allies for their disabled children. In this book, you will read maternal narratives that highlight how mothers resist external ableist structures, whilst also recognising that these mothers often simultaneously internalise and operate within those very same ableist structures. This is not an unusual finding nor limited to this research, rather it is a stark reflection of how many parents perform their parenting roles in ways that observe the rules of professional-parent engagement. For instance, mothers of disabled children in general are used to hearing admiration in awestruck tones from mothers of able-bodied children about how amazing and extraordinary they are because they are mothering a disabled child, and that only they are capable of performing this task. Such admiration frames the mothering of a disabled child as a superhuman trait which is lacking any nuance such as how mothers can experience mixed emotions, frustrations and weaknesses, leaving little or no room for mothers to make mistakes whilst they learn how to mother her disabled child 'on the job'. In fact, any maternal failings risk being viewed by the state as a sign of family dysfunction. This 'supermum' depiction of mothers also dismisses the hardship and trauma that they experience when navigating their child's disability. It could be more complex in Muslim minority contexts where some Muslim mothers operate from a religious perspective that places mothering at an elevated position, suggesting that mothering a disabled child is a gateway to paradise. However, irrespective of disability, Islam values mothering as a noble role to embrace.

Hence, we must understand the multiplicity of worldviews and perspectives that mothers hold when mothering their disabled child. The reason why I am stating this upfront is not to absolve myself as a researcher who could be accused of advancing an arguably ableist narrative, but rather because it is important to acknowledge that parenting in general, and mothering in particular, is not a romanticised role performed within an ideal utopian context. It is often performed amidst contradictory worldviews and scripts, some that foster critical consciousness and some that oppose it. It is imperative to acknowledge the challenges in performing allyship, and how one minoritised oppressed group can be part of a system that reproduces oppression for another minoritised oppressed group. This is required for forging critical alliances if we are truly committed to social justice. I explore this aspect in greater depth in Chap. 4.

In reading the narratives of mothers in my study, I recognise the presence of ableism and patriarchy as much as other forms of oppression;

minoritised mothers of disabled children are tasked with the huge burden of coming across as the right kind of advocate to a Western audience. The Eurocentric colonial perspective around parental advocacy for children with disability, entails taking an individualistic standpoint as the dominant way of executing advocacy; this is one of the factors that disability researchers from the Global South are trying to challenge (Stienstra, 2015). This book, therefore, provides an insight into how minoritised mothers fight their own corner for their disabled children and for themselves and do so in ways that may or may not fit into traditional maternal advocacy roles.

WHAT IS INTERSECTIONALITY?

As this book relies heavily on utilising an intersectional lens, it is important that I provide some context about why this lens was considered. Looking back at my own journey as a student, and later as an early career researcher, I found the course reading lists and optional modules within universities' inclusive education programmes offered very limited perspectives around the inclusion and exclusion experiences of minoritised communities; this is an aspect that I would like to continue to work on in an attempt to develop a more inclusive curriculum for undergraduate and Master's level programmes on inclusion and disability. There is very little acknowledgement of the heterogeneity of the concerns and experiences of British minoritised communities within academic texts. It is only when I became immersed in the lives of my participants as a researcher, that I began to feel alienated from Eurocentric scholarship. I was eager to explore the theoretical influences that spoke deeply to my participants' experiences. At this point, I did not reflect on epistemological racism rather, as Scheurich and Young (1997) suggest, I was beginning to realise that the established ways of researching with ethnic minority communities were not context-free, but were situated in a social history of a dominant White race that exoticises and pathologises ethnic minority experiences. As a result, scholars of colour spend considerable time refuting such problematic depictions of their communities, instead of focusing on more pressing issues and challenges that their communities face (Scheurich & Young, 1997). My engagement with the works of bell hooks and Kimberlé Crenshaw occurred during this time, incidentally in non-academic spaces, initially during discussions with my sister who is Muslim American philosopher and later in other feminist gatherings. I was drawn to the notion of how multiple oppressive structures interact and create specific sites of oppression for minority communities. I could already see that Robina Shah's work, *The Silent Minority* (1995) which was instrumental in challenging perceptions about disabled minority families living in the UK was grounded in intersectionality, even if she had not named it as such. Her work dismantled common misperceptions held by health and social care professionals about South Asian mothers being oppressed by cultural patriarchy, and highlighted how these mothers are not passive participants in their child's upbringing but rather actively make important decisions in running their household.

When I commenced this research project in 2013, I had relative flexibility and time to engage in wider reading outside the education discipline. I was free to read anything and everything that aroused my interest. It is during this time, that I revisited Crenshaw's (1989) work on intersectionality with deeper interest.

> Intersectionality is a way of understanding and analysing the complexity in the world, in people, and in human experiences … When it comes to structural inequality, people's lives and the organisation of power in a given society are better understood as being shaped not by a single axis of social division, be it race or gender or class, but by many axes that work together and influence each other. (Collins & Bilge, 2017, p. 2)

Adopting an intersectional lens—which encompasses the critical facet of social justice—is imperative for researching with minoritised communities. In referring to the powerful quote by Collins and Bilge (2017), I want to emphasise how mainstream disability literature on the experiences of minority mothers does not fully engage with the multiple social inequalities they face, nor does it fully acknowledge that minority mothering takes place in a predominantly White space. Having already completed two different research projects with South Asian mothers of disabled children, I could see that their experiences of raising their children was determined by a complex and simultaneous interaction of their individual positioning with wider social, structural and institutional processes. Whilst the term intersectionality was coined by Crenshaw (1989), intersectionality as a form of critical inquiry and praxis can be traced back to historical figures such as Sojourner Truth's famous speech, *Ain't I a Woman?* and to historical movements such as the Civil Rights Movement, Combahee River Collective, Chicana Feminist Movement and other movements around the world such as the Third World Women's Alliance which challenged the inequalities produced by political and institutional structures (Collins & Bilge, 2017). The advent of Global North universities identifying the origins of intersectionality has more to do with the needs of the university in legitimising its educational disciplines, and less so about the minoritised movements and historical struggles on whose backs intersectionality was created which have been lost and dismissed in the subsequent theorising about intersectionality (Collins & Bilge, 2017).

We have a very sanitised understanding of intersectionality that largely focuses on hybrid identities, but overlooks the multiple oppressive structures that mediates an individual's experiences of social inequalities. The consequence of adopting such a sanitised intersectional lens is that it enables British schools to superficially acknowledge and engage with their minority communities, such as celebrating Muslim holidays like Eid, whilst simultaneously implementing oppressive policies such as Prevent that racially profile the very same communities and continuously evaluate their loyalties to the UK. Such *"racial gesture politics"* (Rollock, 2018, p. 324) are designed to assure minority families that their children's schools value their input, and yet encourages them to situate the cause of their poor schooling and overall experiences to their own selves rather than to schools and broader institutional structures. This sanitised approach also affects how we, as researchers conduct social justice research and our understandable need for minority narratives to be rooted in resistance. Global North universities are only concerned that certain kinds of narratives flourish, and as Pyke (2010) suggests, are focused on restitution or resistance-based narratives. Some may believe that this is not necessarily a bad thing because it projects a positive narrative, however, it provides a partial picture at best, dismissing and in some instances deliberately silencing events, stories and processes that may not strictly conform to a version of intersectional narratives favoured by Global North academia.

> … the inclination to see resistance everywhere and read its many forms "as signs of the ineffectiveness of power and of the resilience and creativity of the human spirit in its refusal to be dominated" is problematic; it discourages certain questions about the workings of power … Most fundamentally, it forecloses attention to complicity, accommodation, and the maintenance and reproduction of domination. The result is the exaggeration of resistance in social life and an underestimation of the power of oppressive structures to limit agency. Consider, for example, that the most subordinated members of society who are the least likely to be able to engage resistance do not, by

definition, rise to positions that permit them to "speak" their experiences of oppression into the scholarly discourse and shape theoretical proclivities. (Pyke, 2010, p. 560)

I want to sit with Pyke's words for a moment because they echo the concerns about researching and writing about minoritised communities in a way that is palatable to a mainstream audience. There is a pressure to write in a way that represents the positive narratives about communities that are overly researched and problematized by mainstream discourse. However, I have not written about resilience here. Rather, I have written the messy truth which acknowledges that oppressive structures are hard to escape, and that they exist in many of our public and private spaces. As I stand on the shoulders of BIPoC feminist scholars who have preceded me, I have learned the harms of being spoken for, the harms of claiming to liberate others, the harms of theorising the trauma of others, whilst ultimately becoming complicit in their oppression. I also understand the irony that whilst this book avoids the script of 'speaking for' or 'giving voice to' minoritised mothers, nonetheless, as a British Pakistani immigrant mother who does not have to navigate the special education system, I am still bringing forward private voices in public spaces for scrutiny and empathy alike—a task that requires continuous ethical engagement and responsibility. I am cautious that by mainstreaming intersectionality, critical consciousness is removed from the concept, leaving intersectionality to only advance the agenda of Global North academia around what social justice research should be like (Collins & Bilge, 2017). It leaves the concept of intersectionality holding little transformative or liberatory power to the oppressed who need it most. By changing its conceptual understanding, intersectionality risks becoming a tool that holds little utility or relevance to those people working at the bottom-most hierarchy. It is very difficult to conceptualise or suggest a single framework for intersectionality because doing so is antithetical to its roots. However, Collins and Bilge (2017) argue that any framework that claims to use intersectionality as an analytical tool must be based on six guiding principles: social inequality, relationality, power, social context, complexity and social justice. These principles have no specific order and nor do they always appear in the same way in different research contexts. Rather, Collins and Bilge argue, they provide a reference point or guidepost for how researchers can think about and approach their analysis so it can be intersectional. These six principles *"reveals how violence* [oppression] *is not only understood and practiced within discrete systems of power, but also how it constitutes a common thread that connects racism, colonialism, patriarchy and nationalism for example"* (Collins & Bilge, 2017, p. 55).

Why Is an Intersectional Lens Needed for This Work?

As feminist of colour disability studies scholars in the USA, Schalk and Kim (2020) argue that there is a danger in ascribing to and centring on *"state sanctioned understandings of disability"* since it effectively rewrites the nation-state as offering protection and recognising the rights of all disabled people, whilst simultaneously silencing the narratives and rights of those that the nation-state has subjected to violence—its Black and Brown communities—through various apparatuses such as immigration policies, incarceration, the school to prison pipeline, and health and housing policies (p. 43). Schalk and Kim (2020) push for the inclusion of religious, cultural and other perspectives that might run counter to a White social model of disability but are significant to the healing and well-being of Black and Brown communities. In line with Schalk and Kim's call to examine state violence and its relation to disabled communities of colour, an intersectional lens allowed me to examine how the experiences of British Pakistani mothers intersect with different state apparatuses and what they see as just and inclusive for their disabled children. This requires examining racism and Islamophobia within disability studies, and ableism and patriarchal structures within religious and South Asian studies. A fixed boundary between disciplines yet again centres on Whiteness and creates its own exclusions that have a significant impact on minoritised communities that experience disability at multiple axes of power. I was drawn to an intersectional lens because it offers a meaningful foundation to develop what disability justice could look like for British South Asian Muslim disabled families. It was important to bring forth certain critical aspects to the mothers' multiple positionalities that are often relegated to a background context, and present these as a main educational issue within the current research. I wanted to discuss religion, immigrant history, gender and culture as the most significant aspects that affected their mothering; these factors could not be relegated to mere reductive barriers that the mothers had to overcome, because doing so would be forcing their experiences to fit into certain institutional narratives around minorities.

Previous literature on minority families' experiences of navigating their child's disability experiences (Chamba et al., 1999; Hatton et al., 2004), whilst seminal to highlighting their poor access to provisions and formal and informal services, does not place mothers at the centre of these familial experiences. They also dismiss the fact that minority mothers are not only aware of their positionings but also that they hold critical consciousness to understand and mediate their situations better. For instance, Hatton et al. (2004) highlighted how disability identification and diagnosis took much longer with South Asian disabled families than with British White families, and that often minoritised parents were excluded from professional meetings. However, Hatton et al. does not consider how South Asian disabled families had simultaneously developed a meaningful framework to understand their child's disability at the intersections of culture, gender, immigrant history and religion. When I interviewed the mothers in my study, it was clear that they were not living their lives in strict binaries and that aspects of their positionality could be empowering in one instance but be oppressive in other situations. In themselves, social categories such as religion, gender, culture and immigration trajectory are neutral, neither good nor bad; however, they impact the experiences of minoritised families and the decisions that they make on a day-to-day basis.

An intersectional research must highlight 'social inequalities' that minorities experience, such as systemic racism and patriarchy that is embedded within institutional and other formal and informal settings. For instance, a majority of British Pakistanis and British Bangladeshis live below the poverty line, and also have lower educational attainment levels, lower employment rates among males and females, poor experiences of uptake of services, and lower rates of house ownership. It is harder to focus on their low levels of social mobility and educational attainment levels, when little attention is paid to how institutional gatekeeping prevents these communities from participating in and exercising their rights to accessing better education, housing, employment and health. It is hard to address one inequality without addressing the others because they are interconnected. Traditional home-school literature within special education should expand its focus in this area. There is also hesitance within special education literature orthodoxy to centre on minority experiences which situates minoritised communities in their own histories and backgrounds as aspects that have sustained these communities through decades of systemic racism, as opposed to merely examining the reasons for their lack of integration and involvement with schools and other institutions.

This call to focus on minority experiences is not new, and has been repeatedly highlighted by Harry and Klinger (2006), Sullivan and Artiles (2011) and other academics exploring the disproportionate representation of ethnic minorities within special education. Yet we find that schools policies, structures, socialising practices, and teacher perceptions and support are still largely based on a deficit model for engaging with minorities. We cannot focus on educational inequalities without examining how they intersect with other social inequalities.

A second principle that is central to conducting intersectional research is examining the various 'domains of power' that mediate professional-family interactions. If we look at literature on various parenting models from the 1980s and 1990s (Mittler & Mittler, 1983; Cunningham & Davis, 1985; Gascoigne, 1995), we can see how these models have categorised parents into types; parents with a certain level of advocacy are more suited to the parent-partnership model, whereas other parents perceived as being passive are deemed more suited to the expert model. This throws up some crucial questions: how do practitioners decide which families are deserving of an equitable partnership, and which families are only there to learn from their professional expertise? How is it that these parenting models do not consider the cultural power of Whiteness within institutions, as well as racism, patriarchy and gender dynamics that affects interactions with parents? Do British White professionals exercise respect and sensitivity when engaging with British Pakistani mothers? Do professionals assume that mothers lack agency and that they consent to someone speaking for them and making decisions on their behalf? What happens when these mothers refuse to accept expert advice? Are their actions understood and admired as advocacy and embedded in resistance, or are they categorised as troublesome and non-integrationists? Are they perceived as not thinking in their child's best interest? Do these parents need to be educated about what is the best possible choice for their child? Underpinning these questions is growing evidence within literature that suggests that certain types of parents of disabled children are recognised as advocates for negotiating on their own terms, whilst others who do not fit the script are portrayed as unwilling and unruly. Is it just White middle-class parents who are recognised as advocates, whilst parents from minoritised groups are not?

If institutions hold a deficit view of minoritised communities, then their interactions and the outcomes of these interactions are likely to put these communities at further disadvantage. As Harry and Klinger (2006) suggest, schools, hospitals and other comparable institutions are more than just places where services are offered, rather they are active sites for reproducing inequalities. The deficit lens was demonstrated recently when it was reported that the Citizens Advice Bureau's (CAB) training guide on interacting with 'BAME communities' contained blatant racism and inappropriate and harmful stereotypes. CAB, a network of 316 independent charities that provides free and confidential advice on legal, consumer and money matters to the general public described minoritised communities as being "*cash-centred cultures*" that enacted "*early marriage and large families*" with "*a cultural focus on honour and shame*" whilst maintaining "*a distrust of British authorities*" (Iftikhar, 2019). Whilst CAB took this page down from their website after it went viral on social media, one wonders whether they have actively taken organisational measures to undo the harm that may have been going on for years.

Annamma (2018) argues that schools assume a certain cultural capital that privileges British White heteronormative middle-class families. Whilst Britain is undoubtedly a multi ethnic and diverse nation, we also have a long way to go before we can be considered a racially just society. Our classrooms still assume that children who come from ethnic minority families lack linguistic capital, even if in reality these children are multilingual or indeed only speak English. Classrooms do not utilise the linguistic knowledge of minoritised families unless it is used tokenistically on school display boards to demonstrate their diversity. This is one of the main principles that underpins an intersectional lens—the domains of power. This principle reveals whose cultural capital has more power within schools, and whose does not. Inside and outside classrooms, how inclusive are educational professionals of minority cultures? A glimpse at the Teachers Standards (DfE, 2011) outlines the Fundamental British Values (FBV) that teachers need to actively incorporate within their classrooms, revealing a more accurate picture of the level of inclusivity within schools. FBV are problematic because they ascribe to a singular political identity of Britishness that is rooted in the values of exclusion and viewing identity differences as problematic (Elton-Chalcraft et al., 2017). More importantly, teacher training programmes do not provide opportunities to trainee teachers to critically unpack and problematize how racism, xenophobia, FBV and ableism intersect and play out within their professional practices. This is compounded by the fact that the media consistently and disproportionately highlights stories and images of minoritised communities as non-integrationist and a threat to British values. It is only when we examine the structural, cultural, disciplinary and interpersonal domains of power that we can begin to judge how a policy such as FBV, that was drafted as a response to the so-called Trojan Horse Affair, has the capacity to systemically single out Muslim communities within educational settings. As Richardson puts it, *"the discourse of politicians and some of the media implies that a central purpose of teaching British values is to control and regulate young Muslims rather than to empower them"* (Richardson, 2015, p. 45).

A third principle that underpins intersectionality is the notion of 'relationality'. Collins and Bilge (2017) posit that relationality is the analysis *"of multiple identities within the interpersonal domains of power"* and requires us *"to understand how class, race, and gender collectively shape global social inequality"* (p. 27). There is a rejection of the binary of an either/or script, rather there is a conscious effort to understand the conflicts that can occur at various intersections. Relationality acknowledges how power works as a function of one's various social identities, whether that is disability, immigration status or cultural and religious practices. A focus on relationality magnifies the tensions that affect the mothering of disabled children. For instance, to what extent can one, as an abled-bodied mother, exercise parental rights in governing a disabled child's sexuality and sexual practices without acknowledging that one's mothering is couched in ableist practices? Relationality is also of particular significance when analysing the dynamics between mainstream British society and the British Pakistani community, and between the British Pakistani able-bodied community and British Pakistani disabled families. How do differences within group identities create experiences of othering for those holding marginalised identities? A number of studies have explored South Asian non-disabled families' experiences within Britain's education system (Bhatti, 1999; Abbas, 2004; Crozier & Davies, 2006; Chanda-Gool, 2006; Din, 2017). Whilst these studies do not examine disability, they do help to contextualise migrant history and how South Asian families have adapted (or otherwise) to mainstream British society. For instance, studies highlight how most British Pakistani and British Bangladeshi families live within ethnically rich neighbourhoods, with strong extended family support.

However, when we examine studies on South Asian disabled families, they reveal an alienated experience with families who do not benefit from strong community ties which could be useful in developing home-school partnerships (Katbamna et al., 2004; Hatton et al., 2004). The lack of meaningful relationality between the dominant group (able-bodied British

Pakistanis) and those with marginalised identities (British Pakistani disabled families), means that the dominant group not only speaks for those at liminal positions without their consent but are also not self-conscious about their privilege as able-bodied individuals and their role in perpetuating ableism within their own community.

A fourth principle that is central to an intersectional research is the need for 'contextualisation'. Whilst attending two separate academic panels in 2019, I was asked if my scholarship belonged in educational research and whether it would be better if I had positioned my research to fit within the field of health and social work. For too long educational research has applied a reductive approach to examining minority experiences in education. For instance, the home-school literature which focuses on special education represents the experiences of minority communities from a White perspective, akin to the example I shared with you at the beginning of this chapter. My response to these academic panels was that if these conversations had not been included previously within the educational research, then their questions only proved to me that these discourses needed to be an integral part of mainstream educational issues.

Our insistence on having clear lines of demarcation between different disciplines means that we will always have an incomplete understanding of ethnic minority experiences at different intersections. Educational research orthodoxy seemingly only recognises the existence of ethnic minority issues as long as they are centred on language and cultural barriers—this is essentialist and misses out on the 'complexity' of multiple social inequalities affecting minority educational experiences, nor does it frame the educational issue as an indication of wider 'social justice' issues. These two guiding principles are crucial to understanding the silencing that minority parents face the moment they speak out about either structural, cultural or institutional injustice, and how their concerns are dismissed as being 'not educational'. This begs the question: whose concerns can or should be counted as an 'educational' issue? What kind of analysis are researchers permitted to use in highlighting the experiences of South Asian disabled families? This is obviously important to address if we are serious about decolonising research, and shifting from academic scripts that have created a deficit view of minority communities. It is important for academics who research with minoritised communities to contextualise their work against the backdrop of austerity and the cuts to funding by the UK government for education, health and social care services that impact minoritised communities disproportionately. This examination should also include consideration of rising xenophobia and racism to develop a deeper under-standing of why certain disabled families are more excluded than others.

Existing research has examined these social divisions strictly in binary terms, meaning that South Asian culture is either portrayed as a wholly positive example of a close-knit community that provides informal support networks, or alternatively is depicted as a purveyor of patriarchy which purposely opposes integration into mainstream British society. Intersectionality also considers the interpersonal domain of power, which means how individuals relate to one another determines the extent to which they are able to exercise agency within that situation. They can have more privilege in one instance and feel disadvantaged in another. This is a key consideration when understanding the dynamics between minoritised mothers and professionals. There is also the question of who holds the most power in perpetuating the othering of South Asian disabled families. Is it the local community centres that act as community gatekeepers, or the mainstream schools that exclude disabled children who occupy an eth-nic, racialised or religious identity? Or both? By engaging in intersectional analysis which explores social inequality, relationality, power dynamics, social context, complexity and social justice, I hope to provide a deeper insight into different and intersecting mechanisms and factors that are at play as South Asian Muslim mothers navigate the UK special educa-tion system.

Why the focus on South Asian Muslim disabled families? What differen-tiates the experiences of South Asian Muslim disabled families compared to the experiences of other disabled families? After all, could this study not be about any disabled family? When you utilise intersectionality as a con-ceptual framework, you can begin to realise why the South Asian Muslim community in general faces a unique set of issues. When you study with South Asian Muslim disabled families you cannot understand their experi-ences without an intersectional lens, as this book aims to demonstrate.

What Are the Main Issues Addressed in This Book?

This book reports on an in-depth qualitative research examining the expe-riences of British Pakistani mothers of supporting their disabled children. Initially, I tried to include both British Bangladeshi and British Pakistani mothers in my research, however, as a British Pakistani I was less successful in establishing contacts and networks within the British Bangladeshi com-munity; perhaps my outsider status as British Pakistani might have affected this access. The mothers in this study were based in the Southwest of England, a region with 91.8 per cent White British population compared to a British South Asian population of 2.0 per cent which is the lowest compared to other regions across the UK. This book does not represent experiences of all British Pakistani mothers of disabled children, rather it aims to demonstrate just the opposite—that the experiences of British Pakistani mothers cannot be homogenised, and that there is nuance in how they make sense of raising their disabled child at various intersections. This book does not create a new parenting model nor proposes new mothering categories for mothers to fit into as you might find in traditional home-school literature. Rather, this book asks professionals to let go of the patriarchal neoliberal categorisations of parents, specifically mothers, in the hope that they provide a clear blueprint of what their professional relationships will be like with minoritised mothers. I suggest a new way of thinking about minoritised mothers without essentialising them and their experiences of raising their disabled children.

The book is divided into two parts. Part I unpacks the need for intersectional analysis and focuses on the minority maternal narratives detailed in Chaps. 1 and 2. Part II examines the intersections of religion, gender, culture and immigration pathways that mediate the experiences of British Pakistani mothers, which comprises Chaps. 3, 4, 5 and 6. When I started writing this book, I considered whether these contexts should be addressed within separate chapters due to the danger of these contexts being incorrectly perceived as fixed binaries rather than as intersecting contexts. However, I ultimately decided to layout each context within individual chapters primarily because these very contexts are so often overlooked within existing literature, or discussed in a limiting way. For instance, religion is often explored within literature insofar as the parental religious understanding of disability and whether it is compatible with different models of disability; however, it is rarely examined in relation to how it shapes Muslim mothering and caring roles for their disabled children. In the course of writing this book, I have also considered how the maternal narratives within each chapter reflect the simultaneous interplay between these different contexts—for instance, looking at how one context, cultural patriarchy places a burden on minoritised mothers in relation to another context, performing gendered care.

Chapter 2 provides rich narratives of eight British Pakistani mothers—Parveen, Saira, Kiran, Tahira, Shehnaz, Maria, Alina and Maham. The main thrust of the Chap. 2 is to lay out these maternal stories about mothering their disabled children as part of a minority in the UK, acknowledging those experiences which have previously been ignored within traditional literature examining home-school and parent-professional collaboration, and in the process channelling private anguish into public discourse. Through these maternal narratives, I ask readers to consider how professionals working with minoritised mothers can work in a more holistic way, looking at individual family needs as well as the ways in which maternal and family knowledge can be utilised to better target the provisions and overall experiences of disabled children. The narratives also unpack the problem of limiting inclusion discourse to merely mainstream school placements, and illustrates that the inclusion and exclusion experiences of minority families are tied to other discourses such as visible and invisible disabilities, a familiarity with navigating Britain's special education system, and the asymmetric power dynamics between minoritised mothers and educational, medical and social care professionals. I discuss how maternal agency is constructed at the various intersections of ethnicity, culture, immigration trajectory, personal family dynamics, as well as the difficulties that these minority mothers experience when professionals question whether they know their child's needs. These maternal narratives also ask readers to examine who the gatekeepers of inclusion and exclusion are in each instance. It pushes us to think how local authorities, medical, educational and social care professionals, community leaders and family members can determine the level of inclusion experienced by minoritised disabled children, young people and their families.

Chapter 3 considers how mothers address the Islamic rights and obligations of their disabled children, whether they utilise a religious framework in understanding their own roles and responsibilities, and how disability mediates the extent to which these disabled families experience inclusion within community religious events. In this chapter, we also consider how maternal religious beliefs do not impede mothers from seeking medical or educational interventions. Rather, religion provides mothers with an alternative lens to the traditional Western grieving model, shifting their focus from coping to becoming a frame of reference for navigating the special education system. I also consider how a greater knowledge of Islamic jurisprudence equips these mothers to challenge cultural practices within the community such as cousin marriages. This chapter discusses the positive role that religion can play towards challenging religious exclusion, and towards developing religious identities that support the gendered nature of care of their disabled children.

In Chap. 4, we turn our attention to whether feminist notions of care conflict with maternal roles as allies and with Islamic definitions of mothering. I also present how the current discourse on consanguinity within British Pakistani communities impacts mothers in the current study; the institutional and community gaze to produce perfect babies means that mothers are judged by others, and that their actions can be perceived as immoral and selfish regardless of whether they choose to bear a disabled child or terminate their pregnancy (Sharp & Earle, 2002). I also explore how British Muslim women's reproductive practices have become sites of scrutiny due to the cultural practice of cousin marriage, in terms of their 'efficiency' in expanding and strengthening their diasporic communities. Chapter 4 also examines the tensions with how religious and cultural beliefs affect the gendered and sexual experiences of disabled children. The chapter highlights the aspirations that mothers hold with regard to equal access to education and employment for their disabled child regardless of their gender. I also focus on how maternal attitudes towards the marriage prospects of their disabled children are deeply entwined with mothers' roles as allies. The maternal narratives reveal how every mother rejects marriage as a solution to managed long-term care for their disabled child.

In Chap. 5, I question whether mothers exercise false consciousness when they operate from a 'religion versus culture' paradigm, and whether there is a monolithic British Pakistani culture to which mothers can relate. This chapter introduces the notion of 'cultural bubbles', and the role they play in how British Pakistani disabled families create their own cultural identities and values based on their own inter-categorical positioning to navigate their daily lived experiences. I demonstrate how new cultural identities are always in the process of being formed, and how they are linked to a child's disability and the family's religious beliefs and priorities. In this way, mothers are able to mediate their inclusion and exclusion experiences from within their local community on their own terms. Chapter 5 also pushes us to think about the role that community centres can play in supporting disabled families, and about whether they need to incorporate more training around maintaining confidentiality and privacy when offering support to families. The maternal narratives reveal that whilst notions of care and inclusion might be enmeshed within traditional collectivist family structures—for instance, grandchildren being cared for by grandparents or the older generation living with and being cared for by their sons and daughters-in-law—it still benefits able-bodied families and requires reframing and reconceptualization before it can truly benefit disabled families. Finally, Chap. 5 critically unpacks why mothers may view institutional spaces—and by extension White people in general—as more professional, inclusive and trustworthy compared to their own community members, and how this affects their engagement with formal and informal services and provisions.

Chapter 6 explores the heterogeneity within the British Pakistani diaspora's immigration pathways, and the way this shapes their experiences as citizens in the UK. This chapter dispels the myth of return that underpins the discussion on migrant families in their host countries. In the context of the UK's hostile immigration policies, I look at who is counted as a valuable citizen and how this is deeply embedded in a White ableist discourse. Chapter 6 poses critical questions around why the needs of ethnic minority disabled communities are viewed as separate to the needs of British White disabled communities within media and official discourse. Who can express grievances against the current state of provisions and interventions for disabled families without being perceived as ungrateful subjects? Which country do mothers refer to as their homeland, and do they feel comfortable enough as citizens to ask for what is rightfully theirs by law? This chapter provides the readers with the stark realisation that regardless of having resided in Britain for decades, ethnic minority disabled families are viewed as 'other' who may never be allowed to feel at home.

This book concludes with recommendations of how institutional and communal spaces can become more inclusive, as well as a discussion on the implications of ignoring the multiple contexts in which the mothering of British Pakistani Muslim disabled children occur. Throughout the writing of this book, I have incorporated as many stories that have been shared with me as possible, and in doing so I have highlighted the different contexts and aspects that shape the mothering of disabled children in the British Pakistani Muslim diaspora. It is my sincere hope that over the course of reading this book, readers will recognise the importance of an intersectional understanding of how minority families support their disabled children in the UK.

REFERENCES

Abbas, T. (2004). *The education of British south Asians: Ethnicity, capital and class structure.* Palgrave Macmillan.

Annamma, S. A. (2018). *The pedagogy of Pathologization: Dis/abled girls of color in the school-prison nexus.* Routledge.

Atwood, E., & López, G. R. (2014). Let's be critically honest: Towards a messier counterstory in critical race theory. *International Journal of Qualitative Studies in Education, 27*(9), 1134–1154.

Bhatti, G. (1999). *Asian children at home and at school: An ethnographic study.* Routledge.

Campbell, J., & Oliver, M. (1996). *Disability politics.* Routledge.

Chamba, R., Ahmad, W., Hirst, M., Lawton, D., & Beresford, B. (1999). *On the edge: Minority ethnic families caring for a severely disabled child.* Joseph Rowntree Foundation, The Policy Press.

Chanda-Gool, S. (2006). *South Asian communities: Catalysts for educational change.* Trentham Books.

Collins, P. H., & Bilge, S. (2017). *Intersectionality.* Polity Press.

Crenshaw, K. (1989). Demarginalizing the intersection of race and sex: A black feminist critique of antidiscrimination doctrine. *Feminist Theory and Antiracist Politics, The University of Chicago Legal Forum, 140*, 139–167.

Crozier, G., & Davies, J. (2006). Family matters: A discussion of the Bangladeshi and Pakistani extended family and community in supporting the children's education. *The Sociological Review, 54*(4), 678–695.

Cunningham, C., & Davis, H. (1985). *Working with parents: Frameworks for collaboration.* Open University Press.

Department for Education. (2011). *Teachers' standards: Guidance for school leaders, school staff and governing bodies.* Department of Education.

Din, S. (2017). *Muslim mothers and their children's schooling.* Trentham Books, UCL IOE Press.

Elton-Chalcraft, S., Lander, V., Revell, L., Warner, D., & Whitworth, L. (2017). To promote, or not to promote fundamental British values? Teachers' standards, diversity and teacher education. *British Educational Research Journal, 43*(1), 29–48.

Fisher, P., & Goodley, D. (2007). The linear medical model of disability: Mothers of disabled babies resist with counter-narratives. *Sociology of Health & Illness, 29*(1), 66–81.

Fylling, I., & Sandvin, J. T. (1999). The role of parents in special education: The notion of partnership revised. *European Journal of Special Needs Education, 14*(2), 144–157.

Gascoigne, E. (1995). *Working with parents as partners in SEN.* David Fulton Publishers.

Harry, B., & Klinger, J. K. (2006). *Why are so many minority children in special education?: Understanding race and disability in schools.* Teachers College Press.

Hatton, C., Akram, Y., Shah, R., Robertson, J., & Emerson, E. (2004). *Supporting south Asian families with a child with severe disabilities*. Jessica Kingsley Publishers.

Hussain, Y., Atkin, K., & Ahmad, W. (2002). *South Asian disabled young people and their families*. Joseph Rowntree Foundation, The Policy Press.

Hyde, M., Punch, R., & Komesaroff, L. (2010). Coming to a decision about cochlear implantation: Parents making choices for their deaf children. *Journal of Deaf Studies and Deaf Education, 15*(2), 162–178.

Iftikhar, F. (2019). I'm glad I called out Citizens Advice's racist training—But it's just the tip of the charity sector iceberg. *Gal-dem*. Retrieved August 16, 2016, from https://gal-dem.com/citizens-advices-racist-training-is-just-the-tip-of-the-iceberg-in-the-charity-sector/

Johnston, M. (1996). Models of disability. *Physiotherapy Theory and Practice, 12*(3), 131–141.

Katbamna, S., Ahmad, W., Bhakta, P., Baker, R., & Parker, G. (2004). Do they look after their own? Informal support for south Asian carers. *Health and Social Care in the Community, 12*(5), 398–406.

Knight, K. (2013). The changing face of the 'good mother': Trends in research into families with a child with intellectual disability, and some concerns. *Disability & Society, 28*(5), 660–673.

Kubler-Ross, E. (1969). *On death and dying*. The Macmillan Company.

Landsman, G. (2005). Mothers and models of disability. *Journal of Medical Humanities, 26*(2–3), 121–139.

McCloskey, E. (2010). What do I know? Parental positioning in special education. *International Journal of Special Education, 25*(1), 162–170.

Middleton, L. (1999). *Disabled children: Challenging social exclusion*. Blackwell.

Mittler, P., & Mittler, H. (1983). Partnership with parents: An overview. In *Parents, professionals and mentally handicapped people: Approaches to partnership* (pp. 8–43). Croom Helm.

Morris, J. (2001). Impairment and disability: Constructing an ethics of care that promotes human rights. *Hypatia, 16*(4), 1–16.

Pyke, K. D. (2010). What is internalized racial oppression and why don't we study it? Acknowledging racism's hidden injuries. *Sociological Perspectives, 53*(4), 551–572.

Richardson, R. (2015). British values and British identity: Muddles, mixtures, and ways ahead. *London Review of Education, 13*(2), 37–48.

Rizvi, S. (2015). Exploring British Pakistani mothers' perception of their child with disability: Insights from a UK context. *Journal of Research in Special Educational Needs*. https://doi.org/10.1111/1471-3802.12111

Rizvi, S., & Limbrick, P. (2015). Provision for learners with SLD/PMLD from ethnic minority families. In P. Lacey, R. Ashdown, P. Jones, H. Lawson, & M. Pipe (Eds.), *Routledge companion to severe, profound and multiple learning difficulties*. Routledge.

Rogers, C. (2011). Mothering and intellectual disability: Partnership rhetoric? *British Journal of Sociology of Education, 32*(4), 563–581.

Rollock, N. (2018). The heart of whiteness: Racial gesture politics, equity and higher education. In J. Arday & H. S. Mirza (Eds.), *Dismantling race in higher education* (pp. 313–330). Palgrave Macmillan.

Ryan, S., & Runswick-Cole, K. (2008). Repositioning mother: Mothers, disabled children and disability studies. *Disability & Society, 23*(3), 199–210.

Ryan, S., & Runswick-Cole, K. (2009). From advocate to activist? Mapping the experiences of mothers of children on the autism spectrum. *Journal of Applied Research in Intellectual Disabilities, 22*(1), 43–53.

Schalk, S., & Kim, J. B. (2020). Integrating race, transforming feminist disability studies. *Signs: Journal of Women in Culture and Society, 46*(1), 31–55.

Scheurich, J. J., & Young, M. D. (1997). Coloring epistemologies: Are our research epistemologies racially biased? *Educational Researcher, 26*(4), 4–16.

Scuro, J. (2017). *Addressing Ableism: Philosophical questions via disability studies.* Lexington Books.

Shah, R. (1995). *The silent minority-children with disabilities in Asian families.* National Children's Bureau.

Sharp, K., & Earle, S. (2002). Feminism, abortion and disability: Irreconcilable differences? *Disability & Society, 17*(2), 137–145.

Sousa, A. C. (2011). From refrigerator mothers to warrior-heroes: The cultural identity transformation of mothers raising children with intellectual disabilities. *Symbolic Interaction, 34*(2), 220–243.

Stienstra, D. (2015). Trumping all? Disability and girlhood studies. *Girlhood Studies, 8*(2), 54–70.

Stoner, J. B., & Angell, M. E. (2006). Parent perspectives on role engagement: An investigation of parents of children with ASD and their self-reported roles with education professionals. *Focus on autism and other developmental disabilities, 21*(3), 177–189.

Sullivan, A. L., & Artiles, A. J. (2011). Theorizing racial inequity in special education: Applying structural inequity theory to disproportionality. *Urban Education, 46*(6), 1526–1552.

CHAPTER 2

Mothering from the Periphery

PARVEEN

Parveen came to Britain 17 years ago from a village in Gujarat district in Punjab, Pakistan. She comes from an educated family; her parents opened the first girls' school in their village. She became the first woman in her village to attain a college degree and actively mentored village girls. She had an arranged marriage to her cousin, who is British-born and ran a family business. During her early marital years in Britain, her in-laws subjected her to physical and mental abuse and ultimately expelled her alongside her husband and baby from their home. She was shamed within the local Pakistani community by her in-laws, and had little fall-back support. Despite these initially difficult years, Parveen has given her family a fresh start away from community and family pressure. Parveen currently has four children, one boy and three girls ranging from her 16-year-old son, Amjad to the youngest, six-year-old Amber who has PMLD, is Statemented and attends a mainstream school. Parveen also had another daughter, Sehr, who passed away six years ago when she was 10; she had similar PMLD to Amber but attended a special school. While our conversations primarily centred on Amber's current mainstream schooling experiences, Parveen occasionally discussed her experiences with Sehr, comparing her experiences of special and mainstream schools.

Our first interview was in Parveen's home. Her sitting room was stacked with religious books, and some yoga mats in one corner. Parveen immediately informed me she was a religious preacher by profession, offering religious workshops to women in her community on various domestic

S. Rizvi, *Undoing Whiteness in Disability Studies*, https://doi.org/10.1007/978-3-030-79573-3_2

issues. I started the interview by asking her about her experiences since Amber was born; instead, Parveen started discussing Sehr, about her birth, her frequent hospitalisations and then moved onto the difficult period around Sehr's death. This was the first time that she used religious allegory.

> Life can be challenging but I've found support in my faith. The way I see it, when a gardener tends his garden he must thin and trim the plants, so if those plants called him cruel, arguing for their place in the garden they only see their own situation, not what the whole garden needs ... only the gardener can view the whole garden, and he'd never want his garden to become ruined. I think of God's earth as a beautiful garden and He'd never want His garden damaged.—Parveen

Discussing Sehr's final days, Parveen was not bitter and suggested that Sehr's death had been *"inevitable considering her prognosis"*; she had accepted the expert medical opinion and made her peace in a religious sense, believing her daughter's death served a higher purpose. She also reflected how her unfamiliarity with Britain's health and social care system as a recent immigrant during that period affected Sehr's diagnosis.

> I still laugh at this. When Sehr was diagnosed, I saw all these professionals in the meeting, so being worried I asked them do we have to pay all of you? I was from Pakistan, how would I know how the UK system works? I hadn't had any kids at hospital before ... if even one of them [professionals] had been condescending, I'd probably have never asked another question again.—Parveen

Parveen's difficult early immigrant experiences were invaluable in teaching her how to equip herself and support Sehr better.

> ... there was a big gap in my understanding ... I remember once Sehr was on medication which she needed on alternate days, but I was giving her daily, that really opened my eyes. I'd done an IELTS [English] course, so grammatically I knew English but not for everyday use ... even if you're fluent in English, medical jargon or names of medicines are difficult to understand. I had to rote learn each [medicine].—Parveen

Parveen explained how having a child with disabilities highlighted her own inadequacies as a mother, making her more determined to utilise all available resources for Sehr. However, even after taking English classes and despite using interpreters, Parveen expressed how she struggled to understand health professionals because they gatekept key information regarding Sehr's condition. This restricted how she could help Sehr and she admitted that she initially followed medical advice on everything, such as keeping her daughter indoors for fear of making her sick.

When I asked Parveen's thoughts about Sehr's disability, I expected a religious explanation; various South Asian family studies[1] examining religious explanations of disability have reported participants describing their children as a 'gift from God'. However, Parveen explained Sehr's medical diagnosis and the years that doctors had taken to diagnose her.

> I researched the odds [of being born with XYZ[2]] were one in 50,000, but I don't think such a severe case had occurred ... now, in the Southwest there are three children including Amber.—Parveen

Therefore, whilst Parveen held a religious epistemology on life she also held a medical explanation for disability, explaining during later interviews how Amber's environment was adapted to support the severity of her needs.

> With a child with any physical or mental function which isn't normal, everything should be adjusted to that child's needs. You must adapt the system around the child.—Parveen

Her responses reflect that she did not live her life in strict binaries, or hold contrasting explanations of disability at different points in her life. It also suggested her religious faith did not impinge on her understanding of her child's disability. Parveen's religious views in relation to Sehr's bereavement do not reflect a natural response to coping with loss, or an absence of medical reasoning in contrast to the findings of Bywaters et al. (2003). Interestingly, Parveen is not a non-expert[3] on religion; rather her position as a religious preacher differentiates her from the other mothers in this

---

[1] Croot et al. (2008) and Bywaters et al. (2003).

[2] I am not stating the precise condition because Parveen's daughters, Sehr and Amber had/have a very rare disease which could easily identify my participant and her daughters, thereby breaching their confidentiality.

[3] Non-expert is a term used by Ammerman (2007) to distinguish between institutional representation of religion from the everyday religious experiences and understandings of lay persons.

study, allowing her to provide a meaningful framework through which she can view her difficult experiences without being perceived as irrational or in denial. Religious sermons by South Asian Muslim male preachers often allude to a detached, patriarchal and often inaccurate explanation of disability, such as labelling disability as a divine gift or punishment, or blaming a mother's supposed sins for her child's disability. Feminist interpretations do challenge these patriarchal viewpoints, however, they do not experience proper uptake in most communities and hence lack due impact. Parveen's personal experiences of supporting her disabled children are based on her theological understanding, providing a more nuanced and authentic view.

Parveen promoted a sense of normality for her disabled children which she acknowledged was challenging, nonetheless, it was necessary to allow them to reach their full potential. She felt this was possible with Sehr's special school, describing her relationship as *"non-intimidating and cooperative"*.

> The staff were really nice, I never felt scared … they always met me with such politeness, so I was at ease.—Parveen

Sehr's special school were experienced in caring for children with PMLD with medical complications, and were understanding about Sehr's absences or medicine supplies. However, Parveen felt Amber's mainstream school were not inclusive and lacked the requisite experience of supporting a disabled child.

> This school is academically one of the top schools in the Southwest … top schools will often try not to enrol disabled children because it dips their academic performance. When schools record attendance, and Amber is absent a lot [because] she has medical appointments, but school statistics don't show the reasons for the absence. The staff are brilliant but they don't understand Amber's needs. I know they're trying their best, but I'd like the school to bring such children towards a normal routine.—Parveen

Parveen felt dissatisfied because Amber's mainstream school had been unable to accommodate her daughter's needs, leaving both Parveen and Amber still adjusting to the school's expectations. For instance, often when Amber had appointments outside school, the school expected Parveen to pick up and drop Amber; this was exhausting and something

Parveen had not experienced with Sehr's special school. Parveen understood that Amber's school were following procedure and not intentionally making her life difficult, nonetheless, these checklist measures were non-conducive to Amber's full inclusion.

> ... they [school] ticked the entire checklist for inclusion on paper, like disabled parking, but not in reality.—Parveen

Parveen also highlighted other issues affecting her relationship with Amber's school. For instance, whilst she met Amber's carers every day, she only rarely interacted with her teacher which affected her understanding about how Amber was doing at school; Parveen suggested various hierarchical communication hurdles needed to be cleared before she could actually approach Amber's teacher.

> ... there is communication, but I haven't been able to make them understand her medical and physical needs. Who should I explain to? The three carers, the teacher, or the SENCO?—Parveen

Poor internal staff communication often had serious consequences. For instance, Parveen highlighted how the school had stopped Amber from entering school because her medicines had finished, classifying her 'at risk'. When Parveen asked why the school had not informed her earlier that Amber's medicines were running low the receptionist replied this was not her job.

> I'm always scared about what this school might say to me, whether I've missed anything. They're polite but whenever I've suggested in her diary about Amber using the electric wheelchair or the vest she should wear, they don't do it. Maybe because I'm not a professional, I'm taken less seriously.—Parveen

Parveen suggested that she only met Amber's teacher when she had been unable to provide something for Amber; consequently, she dreaded going into Amber's school, contrasting starkly with Sehr's school. Parveen often turned to her Disability Service nurse, a professional who she considered the school would take more seriously.

> When her [medicine] expired, they sent Amber home with me. I was so angry, I didn't want to send Amber to that school anymore ... ordering the

new medicines could've take up to four weeks, so she shouldn't attend school for four weeks? Disability Services helped mediate, came to my place, provided the medicine, resolved this whole situation. So now there's a middleman I approach whenever I feel stuck.—Parveen

Even with this *"middleman"*, Parveen admitted that her relationship with Amber's school had deteriorated, and she could not continue like this for another five years. She claimed the school often used Amber's Statement to exert power. For instance, the school immediately rejected any suggestions Parveen made regarding Amber, suggesting they could not deviate from Amber's Statement making it difficult for Amber to access classroom activities tailored to her needs. In retrospect, Parveen regretted not appreciating the value of Statements at the time, or how schools adhere to Statements as a strict guideline.

[The school] said to refer to the Statement as a bible, but I had no idea about its value. After I started questioning them about the support Amber should get, they'd refer me back to the Statement and say you read it and approved it.—Parveen

Parveen eventually enrolled Amber into Sehr's old special school. Her experience with one mainstream school convinced her that mainstream schools generally would not accommodate disabled children, placing sole responsibility of supporting the child on parents. Parveen's experiences of mainstream school are not unique, as we will learn from Shehnaz's narrative. She felt the school did not value her contributions, suggesting that the staff and the physical building were *"unfit to accommodate Amber's needs"*. Sehr's special school made Parveen feel included, where all stakeholders shared responsibility in working towards Sehr's goals. Parveen was distressed by how events transpired with Amber's mainstream school, which she had jointly chosen with Amber's medical practitioners. Parveen's narrative reflects that maternal expectations and their relationships with schools may depend on how they 'fit' within, and whether they 'feel' welcomed by, school policies, infrastructure and attitudes.

Parveen revealed why she initially chose this mainstream school,

[Amber's] school and Sehr's school are next to each other. I assumed they were collaborating because when Sehr was alive she'd go [to the mainstream school] once a week, children were going from special to mainstream and vice versa so they could integrate. That gave the impression both schools were linked. The professionals who recommended the mainstream school thought the same.—Parveen

Parveen had assumed their close proximity meant that Sehr's special school and Amber's mainstream school collaborated, and that mainstream staff were trained in working with disabled children. Unfortunately, Parveen only realised after Amber's admission that this was not the case. This affected Amber's support because she needed to use the special school's hydrotherapy pool, which her mainstream school did not have formal access to. Parveen also thought that sharing a classroom with able-bodied children may help create a sense of normality for Amber. She suggested that the disadvantages of a special school are how the whole building communicates 'disability', from the special access doors, wheelchairs and medical equipment on open display, to the specially adapted shower rooms. Therefore, despite her positive relationship with Sehr's school Parveen wanted a fresh approach for Amber, choosing a mainstream school where 'ability' was plainly evident. However, her demanding relationship with Amber's school affected her caring responsibilities at home, making real collaboration less likely.

If they [suggest input] it shows someone else is also concerned about my child, so if I'm struggling alone to support my child, I'd appreciate if someone supports me. That's missing.—Parveen

In our final interview, Parveen was waiting on Amber's placement to be confirmed at Sehr's special school. Despite her disappointing experiences with Amber's mainstream school, Parveen was vocal in expressing her contentment that Amber's health and social care needs were being fulfilled. She appreciated the unrivalled care both Sehr and Amber had received from Britain's health and social care system.

Amber received the same surgery as Sehr. It's very expensive but she had three surgeries. I think if I sold my house ten times over, it still wouldn't be enough to pay [medical expenses]. I wouldn't have been able to provide the care my children received here [Britain].—Parveen

Interestingly, as an immigrant mother Parveen has rationalised her poor experiences with Amber's school as just a negative aspect of life in Britain.

She suggested that the primary healthcare that Amber received overrode any negative experiences, because *"everything else takes a backseat"*. The excellent medical care her daughters had received, she felt, obligated her to contribute to British society.

> I've started volunteering in elderly care homes, cooking for them once a week ... During last year's flooding, I raised £10,000 with my friends and went to flooded areas to help ... my local community were surprised I didn't send this money to help the poor in Pakistan, but I told them this country needs you, you live here so they have a right to expect your support. This is real Islamic teaching.—Parveen

Parveen realised that she could represent an 'ideal Muslim' to British society, enhancing their opinion of Muslims. Moreover, working with professionals from different ethnicities helped her embrace and work with other ethnic groups, ensuring her social network was not restricted to British Pakistanis. She suggested this enriched her experiences of supporting her daughters, and felt upset that many South Asian families did not embrace 'British' values. She described such families as *"culturally religious"*, enforcing a narrow-minded and spuriously religious mentality such as refusing formal services, because they were offered by non-Muslims. She believed this pseudo-Islamic mindset existed primarily to maintain patriarchy, which was detrimental to the experiences of many South Asian mothers.

> ... some people have a totally un-Islamic mentality. A friend said, I can't let my daughter go to respite services because my husband said they're non-Muslims, we don't trust them. I said, then take your kids out of school as well! Stop eating canned food because it was probably packed by non-Muslims! What type of philosophy is that?—Parveen

Parveen suggested her deeper theological understanding allowed her to live more authentically and successfully as a Muslim in Britain than other British Muslim mothers. As a religious preacher, she questioned how her Pakistani friend's husband could misrepresent religion to promote a culturally gendered view of caregiving. Within Pakistani culture, all domestic functions are still viewed as a woman's responsibility which sustains patriarchy, reinforcing the power of heteronormative religious men. In rejecting respite services because it was *"un-Islamic"*, her friend's husband formalised caregiving as his wife's domestic responsibility. Parveen states that such practices are detrimental to women who, as primary carers of their disabled children, desperately need of all forms of support. She also suggested that her reflexivity enriched her experiences of supporting her daughters, compared to those British Pakistani Muslim families who still maintained problematic aspects of Pakistani culture.

> Religion, culture and British society, it's a package, everything is related. Divisions make life difficult, like if you're saying this is where religion ends and where culture starts, it's confusing. Whatever good there is, take it, and whatever doesn't make sense, leave it. Parents get affected by such factors, but I've realised that when I ignore these social divisions, they don't affect me.—Parveen

Parveen suggested that caring for disabled children becomes more challenging if you are worrying about instilling cultural or religious values. She reasoned that parents should focus primarily on supporting their children, ensuring they have the best possible outcomes. She also criticised community members who compartmentalise their Muslim and British identities. As a religious preacher, faith was significant to her identity, however, she felt this should not obscure her other roles. Parveen suggested that her daughters would have lost out on critical provisions had she been preoccupied with cultural patriarchy. This diverges from the dominant narrative surrounding South Asian disabled families, of rejecting formal services if they are deemed culturally or religiously inappropriate (Shah, 1995; Chamba et al., 1999; Croot et al., 2008).

Parveen explained how her efforts to ensure better support for her daughters had been entirely her own initiative. She reflected how mothers often waited to secure the right support, which was risky when the UK government has cut funding across services; accessing support required greater awareness of available supports systems. After Amber's unfavourable school experiences, Parveen preferred to act on her own accord rather than trust the school to change their approach.

SAIRA

Saira is British-born, with familial roots in Mirpur-Azad Kashmir. She mentioned that she was not born into an education-oriented family; she attended mainstream school until Year Ten before getting married. Saira's early marriage to her cousin from Mirpur-Azad Kashmir was non-consensual. During her difficult early marital years, with her husband settling in Britain, Saira and her husband were supported by her family and the community. Saira has three daughters and two sons; the eldest is 20-year-old Shania who attends college, and the youngest is five-year-old Zara who attends reception class. Three children, 19-year-old Faraz, 15-year-old Farha and Zara have mental health needs. Faraz, who has ADHD and was Statemented, has been excluded from nearly every mainstream and special school that he has attended. Farha has anxiety and depression, and had recently been removed from School Action Plus and given occasional SEN support. Zara has attention issues and receives SEN support at a mainstream school. Interestingly, whilst Faraz no longer attends school, Saira discussed his experiences the most during interviews, particularly her difficulties in building relationships with different schools.

I contacted Saira through Anokha, a local South Asian community centre. As a mother of children with behavioural needs, her narrative highlights how such mothers are viewed less favourably by professionals and within existing literature.

We met in a private room in a community centre at Saira's request because she did not want interviews conducted at her home. In our first interview, Saira expressed relief that she could speak openly. She explained that although her pregnancy with Zara had been tough, she was born healthy with no visible difficulties. She initially attributed Zara's hyperactivity and clinginess to being the youngest child in the house. However, both she and the school noticed that Zara needed help.

> Her behaviour [was] quite bad, so she got one-to-one play therapy at school, I knew she needed support but didn't know how much, or how it was affecting her. Farha was going through a bad time but I didn't realise Zara was affected. If I gave her an activity, she'd fiddle between different things, quickly lose her attention. I realised she lacked concentration.—Saira

Farha experienced anxiety and depression in Year Ten at her mainstream school, but Saira did not realise that what was happening to Farha would also have an impact on how much she could support Zara. She confirmed that Zara was currently receiving play therapy and one-to-one support at school, but she avoided labelling Zara's precise difficulty. She suggested that she resolved to approach a family intervention worker who helped secure school support for Zara.

If I hadn't asked, she would've slipped through the system. I met her teacher and family intervention worker and told them her needs.—Saira

Saira contacted a family intervention worker after the school complained about Zara's classroom behaviour. She became worried that if Zara was left unsupported, then she may slip through the British education system.

I'm more knowledgeable now than when Faraz was diagnosed with ADHD. Mainstream school was impossible for him, I let him down because I got him a Statement but I didn't know his rights, whereas with Zara having already been through the process I know my expectations.—Saira

This was the first time she had mentioned Faraz, or used the ADHD label. She explained how she lost trust in the special education system following her experiences with Faraz. She fought to get Faraz a Statement; however, she felt that she had "*let him down*" because she had not utilised it effectively to advocate for his rights. Soon after Faraz's ADHD diagnosis, it became increasingly difficult to justify his mainstream placement to his teachers, making Saira more circumspect and consequently more aware of Zara's rights. As a British-born Pakistani who possessed insider knowledge of the UK schooling system, although language barriers were not an issue, she stated that she faced difficulties in acquiring information and resources for Faraz.

It was so difficult with Faraz but for parents who don't speak English, how many children slip through the net? In our community there's a big language gap ... when you don't know your rights to education and your child has special needs, what support is there? How do you access it?—Saira

Saira also sympathised with those families whose "*children slip through the net*", missing out on support due to the parents' poor English proficiency. She also reflected how a lack of awareness of mental health needs by parents of disabled children within her community, resulted in their children missing out on their entitlements to SEN support. Her reflections on the difficulties in accessing appropriate support resonated with Parveen's experiences; even without language barriers, many families struggle to negotiate the special education system.

Saira also revealed that initially, she uncritically accepted the school's advice regarding Faraz.

> I didn't know what ADHD was. I just told teachers, if that's what you think, I'll go along with it but things went downhill.—Saira

Her lack of knowledge about ADHD meant Saira initially relied on the school's recommendations, often meaning that Faraz was sent home for being 'disruptive'. She recalls interactions with unhelpful and unsympathetic authorities.

> The school got extra money from Faraz's Statement but no provisions were arranged. I didn't know I could say that he's entitled to get an extra teacher, to fight for extra support … they [school] couldn't cope, so he got offered a special school, and I thought that's how it is, I've got no choice. That was in Year Five.—Saira

Saira believes the Statement did not protect Faraz from exclusion or indeed guarantee any provisions; in fact, being Statemented provided his mainstream school with a rationale for recommending Faraz's transfer to a special school. Saira reflected how, at the time, she was unaware of her right to object to Faraz's special school placement, rather than accept it as a fait accompli. Saira suggested Faraz's transfer from his mainstream school magnified his feelings of exclusion, particularly after experiencing racist bullying at his new predominantly White British special school. Unfortunately, teachers dismissed Saira's complaints.

> Faraz felt left out, he was the only Pakistani child and the rest were White. He was repeatedly called 'Paki', which I reported but the teachers said it's not happening. He wasn't learning anything at special school. He wanted to go to mainstream school. The head teacher said he was doing really well, so why couldn't he go back to mainstream school? I was being fobbed off, he was never given the opportunity to prove himself at a mainstream school.—Saira

Saira revealed that Faraz did not cause classroom disruptions in his new special school, and was not sent home. However, he was racially bullied at school and desperately wanted mainstream re-integration, feeling further isolated because all his friends and siblings attended mainstream school. Although the special school initially promised mainstream re-integration,

Saira felt bitterly disappointed that Faraz was not given this opportunity. Faraz's mainstream exclusion and the racism he faced drastically affected his educational experiences. Saira felt that she had failed Faraz through her inability to advocate for him within an education system that had dismissed her concerns, engendering feelings of maternal inadequacy.

> He was referred to CAMHS who prescribed him Ritalin. I wasn't told he could get addicted. I noticed he lost weight, he wasn't eating properly, he just sat there like he was doped out ... afterwards he refused to take it. That was Year Eight. He took Ritalin for five years from mainstream into special. At first, he was on small doses but then it increased. I looked into Ritalin's side-effects, as adults you become addicted, it's powerful drug. Afterwards he was prescribed Concerta. CAMHS just don't want a child who's jumping about.—Saira

Faraz was referred to children and adolescent mental health services (CAMHS) and prescribed Ritalin, that is often prescribed to individuals with ADHD. Saira vehemently maintains that she was not informed of Ritalin's side-effects, and that she only later witnessed its detrimental impact on Faraz. She also suggested that Faraz was aware of professional perceptions that he was merely 'naughty', resulting in his depression over time. He eventually refused to take his medication, after which his special school permanently excluded Faraz. Saira was critical of CAMHS for prioritising classroom management over Faraz's needs. Even as a British-born Pakistani possessing insider knowledge about the education system in general, Saira's experiences of mothering a child with mental health needs reveal that she was still an 'outsider' unless she acquiesced to professional advice. She believes that the school could have found creative ways of teaching Faraz, rather than insisting on medication.

> Faraz walked into the middle of the road into traffic and just stood there. He said I've had enough, I don't want to live anymore. I had to deal with it all by myself, there was no support.—Saira

His exclusion from all types of school environment ultimately led Faraz to become suicidal. Reflecting on this extremely testing period, Saira did not know how to support Faraz and recalls that the local authority did not offer any protection; their failure to include Faraz within suitable education settings damaged his aspirations in adulthood.

Faraz was labelled 'naughty', he wasn't given a chance. Whereas when Zara's gone through a similar phase she's had full support with regular school contact. The most support I've had was from Farha's family intervention worker. Farha's still receiving some support, the most support of any of my kids and it started-off with me referring her myself. Then Zara got picked up [by the same intervention worker]. If I hadn't referred Farha, maybe Zara mightn't have got support so early … I'm not putting Zara on pills.—Saira

Through Faraz's negative experiences, Saira became more informed about her children's educational entitlements. For instance, with Zara, she has not pursued a Statement because it was used against Faraz. She believed that "*Faraz was labelled 'naughty'*" merely because his Statement had mentioned his ADHD, which affected professional attitudes towards him. Saira was adamant that she would not accept putting Zara on medication to manage her concentration, and would be involved in all decision-making regarding Zara's schooling. Faraz's depression also made Saira conscious of how severe Farha's mental health needs were, and whether existing SEN support at her mainstream school was sufficient.

Farha was the youngest before Zara came along, that's when her depression started. She didn't display any needs, she was well-behaved, there were no school complaints but she felt the girls were bitchy. I think she was bullied, so I told the school and they changed her grouping.—Saira

Initially, Saira was slow to recognise Farha's depression due to her hectic schedule after Zara's birth and her involvement with Faraz. Moreover, Saira did not receive complaints from Farha's school until Year Ten, when she started missing school. Saira initially assumed Farha's depression was due to Zara's birth, having previously been the youngest in the household. However, she became worried when Farha started insisting on staying at home from school. Eventually, Farha confided to Saira that she felt excluded and intimidated by the girls in her class. Saira decided not to wait for the school to act, self-referring Farha to a family intervention worker to mediate with Farha's school.

Farha wasn't attending school and they threatened me with court, but I didn't know how to get her into school. The family intervention worker really fought my corner, she understood that Farha's got needs which are now being addressed.—Saira

Without any prior warning, the school threatened Saira with a fine. Saira approached the family intervention worker to convince the school that Farha was being bullied and she was genuinely struggling to feel included. After involving the family intervention worker, Farha started receiving one-to-one support. However, the school had initially dismissed Saira's complaints, and suspected that Farha was absent due to paternal physical abuse. Saira suggested this was not only humiliating but patently untrue.

> They automatically thought there must be something happening at home, looked at me like I'm a bad parent. Of course nothing happened at home [except that] every morning we'd have an argument about her [Farha] going to school.—Saira

The accusation of paternal physical abuse suggested to Saira that the school wanted to shift blame for Farha's mental health issues, and were not serious in addressing them. Saira felt dejected and judged by this experience, relating how she ensured her children attended school, completed their homework, ate healthily and were cared for and yet she was still viewed as an inadequate mother. Moreover, Saira blamed the school's accusation of abuse on existing negative stereotypes about the prevalence of domestic violence within the South Asian community.

> Teachers think, 'they're Asians, when their daughter's grown up she'll leave school'. But I'm encouraging her to attend university, make her own identity.—Saira

Saira highlighted other damaging and stereotypical ways that began to frame the lens through which the school viewed her. For instance, the school assumed that Farha would be taken out of school to get married in Pakistan. She vouched that, as a survivor of non-consensual marriage herself, this was anathema to her. Noting that her eldest daughter was attending college, Saira emphasised that she wanted all her daughters to acquire independence through higher education. Involving her GP and the family intervention worker, Saira faced a daunting task—to dismantle this hurtful stereotype and at the same time ensure that Farha's mental health needs are met by the school. Nonetheless, this episode raised doubts about how Farha's school perceived her and her family. Recognising that the school had problematized Saira damaged her relationship with the school. Such double consciousness that Saira displayed is often experienced by silenced and minoritised families (Du Bois, 1897). Farha's teachers engaged in a reductive categorising of her needs and viewed them as a consequence of paternal physical violence in the home. Inevitably, this ate away at Saira's dignity and subsequently affected her trust in Farha's school.

Saira also expressed how the negative image of Muslims often made her reticent in approaching schools, for fear of being further problematized.

> The PTA at Zara's school is held in the pub … it should be somewhere where it's acceptable to everybody. But I can't be the one saying that, they're already anti-Muslim, I already feel judged. I didn't want to make an issue of it. I just don't go and that's a shame.—Saira

Saira also revealed that she previously attended PTA meetings at Zara's school until meetings were moved to a local pub. As a Muslim, Saira felt uncomfortable going into a pub, however, she refused to seek a change of venue because she *"didn't want to make an issue of it"* in case the increasing anti-Muslim sentiment within Britain magnified this as a Muslim issue. Instead, she simply stopped attending PTA meetings, which she felt was a shame. Saira's passive withdrawal from the PTA reflects how a fear of Islamophobia had become entwined with her identity as a Muslim parent. In order to mitigate her membership of a 'suspect community' (Pantazis & Pemberton, 2009), she chose to compromise with her child's education.

As well as criticising negative media and public attitudes towards British Muslims, Saira also recognised the biases and cultural pressure manifested by the British South Asian community.

> I've always supported my children without judgment. I think my perspective is different to people who've migrated from Pakistan. I grew up here, I understand being in a diverse community, our children are taught two different cultures, doing one thing at home, another outside. They've got two different lives. I accept what's thrown at me and deal with it.—Saira

Saira reflected that as a British-born Pakistani she had experienced the complexities of embracing her Pakistani identity in a White space. She understood what her children were undergoing, and so would not judge them. She viewed her experiences as separate to and perhaps more informed than first-generation immigrant mothers, in that she perceived herself as more embracing of her child's specific needs and dealing with those challenges, rather than denying why her child was experiencing difficulties at school. This was also present in Maria's narrative, who is also a British-born mother in my study; she pointed out the dismissive attitudes of Pakistani-born mothers towards their children's mental health needs. This was interesting perception that British-born mothers held towards first-generation immigrant mothers, thinking that they still needed to develop a greater understanding of their children's special needs as well as hinting that they might be in denial. This internalising of stereotypes reminded me of the narrative I had heard from Alice (in the Chap. 1), and which I explore further in the Chap. 6. It is quite possible that residing in ethnically rich neighbourhoods meant that Saira faced greater pressure to identify with Pakistani culture. This pressure was less for mothers in my study, such as Kiran, who was British-born and lived in a predominantly White British area. Saira's account highlights the complexity of raising children in ethnic neighbourhoods which have their own collective identity, and which seek to distance those who threaten this identity. Saira actively challenges this cultural conformity from within the community to counter the mixed signals it sends to her children, recognising how exclusionary and ableist the community influence is and doing everything to shield her children form it.

As our interviews concluded, Saira suggested that in the process of having these conversations she had become more confident in holding Farha's school to account and to push them to provide the right SEN support. She also felt more hopeful about Farha's school due to greater awareness about how to advocate for her daughters whilst they remained at mainstream school, with the family intervention worker's support.

Kiran

Anokha, who provided Kiran's contact details, suggested that she might be difficult to talk to. However, once Kiran was assured that I did not work for Anokha, she consented to participate. Kiran is a British-born Pakistani Muslim, her parents having migrated from Punjab a few decades ago. Her only child, 15-year-old Ahmed, has global developmental delay. Her husband, a British-born Kenyan-Asian Muslim, is a pharmacist. Kiran works part-time in an administrative capacity in a mainstream nursery. She speaks English and Punjabi. She settled in the Southwest in a single unit household after marriage, having grown up in Northern England.

Our first interview was at Kiran's bungalow on the outskirts of the city, in a predominantly White area. I again assured her that I did not work for Anokha, a community support centre and my gatekeepers, and that whatever she disclosed would not be shared publicly. This assurance satisfied her, and she began from when Ahmed was born, recalling her easy pregnancy,

> My pregnancy was easy and so, you know, for a few years I went through an emotional phase, asking 'why us?' You know, we don't drink, we don't smoke, we don't do drugs ... you know, when you think of those women who do drugs during pregnancy, they're having healthy babies, I went through that emotion. It took me a long time to get over that.—Kiran

Each time Kiran said *"you know"* she looked at me, as if she was asking me to draw on our collective values as British Muslim women who refrain from drinking, smoking, or having boyfriends or girlfriends in our teenage years due to either religious or cultural reasons. She also made an unsolicited confession that she had followed prevailing health advice about abstaining from activities that endanger her unborn child. Kiran's frustration was not unfounded; mother blaming and stigma have been reflected in decades of health policy and awareness campaigns linking alcohol consumption during pregnancy and the prevalence of disability in babies, which is internalised by mothers and manifests as guilt, depression and feelings of inadequacy (Bell et al., 2016). Why should she not feel confused having followed state-sanctioned acceptable behaviour for expectant mothers, only to discover that it was still not enough? The welfare state has treated mothers of children with disability in a patronising, gendered, ableist and self-serving manner. For decades, the state has indoctrinated expectant mothers about acceptable behaviour, such as breastfeeding or vaginal birth that were regarded as 'natural' and thus acceptable mothering; consequently, using formula milk or choosing caesarean births were considered 'unnatural' or negligent behaviour. Therefore, if expectant mothers follow these health guidelines, and if a mother is then informed that her baby is not considered 'typical', it inevitably implies that she must have done something wrong.

Kiran suggested that, as a practicing Muslim, she did not engage in any 'unacceptable' habits regardless of her pregnancy, so following official heath advice assured her that she was, in her words, *"doing the right things"*. After Ahmed was born, she directed her frustration at God and the health system that had promised her a healthy baby. She said that it took a long time to reconcile everything.

I've gotten over that now. I don't think, 'why me?' anymore but it was hard.—Kiran

Perhaps it would be more understandable to equate her emotions as 'grieving the loss of a perfect child', which characterises the traditional discourse of mothering a disabled child; Kiran recognises that her expectations have been influenced by ableist societal standards.

You always want more, to be one step ahead.—Kiran

However, mothering is socialised into ableism through health visitor visits and support group meetings. Kiran recalls that she first noticed that Ahmed was different when she attended mother-toddler groups.

He wasn't doing what other children did, you know, he wasn't interested in toys, the environment, something wasn't quite right. You just know as a parent, don't you?—Kiran

This indicator that Ahmed was developing at a different pace as compared to his age group, that he may be different, was not limited to health or educational settings. As time passed, when Kiran ventured into public spaces with Ahmed, she experienced public hostility and antagonism. Kiran attributed this hostility to the public's ignorance of different types of disability, resulting in differing levels of empathy. Kiran suggested people had to be convinced that Ahmed was just like any other child and that he needed to be treated as such,

You know, with Ahmed, his disabilities aren't visible … you know by his behaviour something isn't quite right. If a child has visible disabilities people say, 'okay, he's got Down Syndrome'. With Ahmed, he doesn't look disabled, so people don't help when we're out. They say, 'oh he's just naughty', but I say, 'he isn't naughty, he just doesn't understand or have awareness'.—Kiran

Notions of invisible and visible disability are significant for mothers whose child "*doesn't look disabled*" and so is often mistaken for a non-disabled child. These mothers are often stared at, and their children publicly chided as if they should be controlled and not be allowed in public spaces if they 'act out'. Griffith and Smith (2005) and Runswick-Cole (2007) suggest that mothers of children with behavioural needs are often blamed for their children's behavioural problems, and expected to 'manage' their home environment better. Kiran exercised constant vigilance and perception management while she was out with Ahmed, constantly informing everyone that Ahmed was not a 'typical' child and in doing so constantly reminding herself that her child was disabled.

> People always look, don't they? We all look at abnormality, it's human nature.—Kiran

Kiran was unconcerned by people staring at Ahmed whenever he vocalised loudly in public, revealing that it was just a part of her experiences of taking Ahmed out. However, she was upset when people were rude towards Ahmed, treating him as if he was fully aware of his actions. She recalled that often in supermarkets, Ahmed wanted to shake hands with the cashier who typically refused and glared at him; it was only when Kiran informed them that this tall fifteen-year-old was a child with special needs and this was how he greeted people that they reluctantly reciprocated his handshake. Kiran found that inclusion was a constant struggle, so settled for tolerance; it was a struggle that she could not experience endlessly, and at times would want to walk away.

> You get tired of always having to explain to other people, this is an ongoing thing with Ahmed … It depends on how you're feeling, what your mood is. Sometimes you're more tolerant, sometimes you're not.—Kiran

For Kiran, the emotional labour of being tolerant and "*always having to explain to other people*" was taxing. Kiran felt that the only place where Ahmed was fully supported and included was his school; his school was a significant aspect of Ahmed's social life. However, special school was not Kiran's first preference. She wanted Ahmed to continue onto a mainstream school because he had attended a mainstream nursery, and because she wanted Ahmed to experience schooling like other children his age. However, working within a mainstream school herself over a number of years, she observed first-hand how mainstream schools went about including disabled children, knowing that Ahmed would be unhappy.

I thought about keeping him in mainstream but realised if he doesn't have any speech, it won't benefit him. He'd be bullied. That's not a good setting for him. It had to be a special school.—Kiran

When I ask her if that was not excluding Ahmed from mainstream schools, she replied,

People say, 'mainstream means inclusion' so we put special needs children into mainstream school … but that child most of the time is taken out of the classroom because they can't cope, so it's not inclusion.—Kiran

Kiran had frequently witnessed children with disability being removed from their classrooms in the mainstream school where she worked, although this had not happened to Ahmed at mainstream nursery. However, knowing this could happen to Ahmed at mainstream school and that he was not able to communicate his experiences, Kiran chose a special school. Kiran's choice of special school might be termed exclusionary by some advocacy groups such as the British Council of Disabled People (BCODP).[4] Attending special school also had its drawbacks; for instance, Ahmed would miss out on studying in a mainstream setting accessing a mainstream curriculum alongside non-disabled children. Kiran also knew that disabled children were more likely to be grouped together without any consideration of how mixed group settings could benefit children in a special school.

Ahmed needs a role model so he can copy what other able-bodied children are doing. With special school, you get all different kinds of special needs children. Some of them will be in wheelchairs, so he wouldn't have that role model.—Kiran

Kiran's experiences of navigating her child's disability experiences illustrate her criticism of superficial locational inclusion, as offered by mainstream schools; however, she also internalised ableism to some degree. Despite the exclusionary attitudes of mainstream settings, nonetheless, she desired the company of able-bodied children for Ahmed. Kiran repeatedly highlighted during our interviews how better provisions, a tailored curriculum and one-to-one support made special school an ideal placement

[4] The BCODP campaigns for all disabled children, irrespective of severity of disability, to be taught within mainstream schools.

setting for Ahmed. She also knew that small class sizes helped Ahmed be more confident at school, which was impossible at mainstream school. This sentiment was shared by many mothers I interviewed. Since 2010, funding cuts to local authorities resulted in specialist provisions being drastically cut from mainstream schools, leaving little support for disabled children at mainstream schools (Ryan, 2019). Consequently, enrolment has increased at special schools; pupil enrolment in maintained special schools increased from 38.2 per cent of all disabled children in 2010 to 44.2 per cent in 2018 (DfE, 2018). However, funding for special schools has not been safeguarded either; Kiran reported that Ahmed's provisions had also been affected by budgetary cuts.

> Speech and language sessions were cut back because of funding. Now he's got a communication aid and uses PECS, does some Makaton ... his Makaton isn't up to it, he can't communicate his needs.—Kiran

Ahmed benefitted enormously from one-to-one speech and language therapy; however, alongside other provisions, this had been gradually phased out due to funding cuts. Ahmed was now using communication aids that only helped him communicate his basic needs, rather than help his speech development. As our interviews came to a close, Kiran expressed her frustration with regard to the gradual cuts to provisions, nonetheless, her own practitioner background sympathised with the school's predicament and helped her to reconcile that schools were doing their best against the backdrop of austerity.

TAHIRA

I contacted Tahira through Anokha, although they knew little about her since she rarely attended Anokha's sessions. When I first met Tahira, I briefly introduced myself and my research. She is a Punjabi Muslim from Faisalabad, Pakistan, where she was schooled until eighth grade. She is fluent in Punjabi and Urdu but not English-proficient, so our interviews were conducted in Urdu. She emigrated to Britain in 2002 after marrying her cousin, a British citizen, and lives in her in-laws' house with her husband and three children, Shehzad, Farrukh and Mona. Tahira is the primary carer to eight-year-old Farrukh who is disabled, to her husband who also has a learning disability, and to her aging parents-in-law. In our first meeting Tahira revealed that her daughter, Amna died when she was one-and-a-half which had affected her other children's emotional wellbeing. Her husband did not work, and their only income was her father-in-law's side-business. I quickly learnt that she was uncomfortable talking with her in-laws present, and scheduled all future interviews when they were absent. Tahira later requested that we reschedule interviews when her in-laws visited Pakistan, delaying my interview schedule by a month; however, I understood Tahira's concern for privacy. Tahira did not describe Farrukh's disability; however, from our interviews and meeting Farrukh briefly, I learnt he was non-verbal and had a metabolic condition requiring frequent hospitalisations. He was Statemented and attended a special school and a disability service for children with life-limiting conditions. Tahira's family dynamic dictated that Farrukh's schooling decisions were made jointly with her parents-in-law.

I first met Tahira at the home of her parents-in-law in a predominantly Pakistani neighbourhood. She was already eager to participate even before I had fully explained my research to her, which was unusual because she spoke little during an hour-long conversation. Her in-laws seemed suspicious of my presence, lingering in the kitchen where Tahira and I were sat. As I was leaving, Tahira followed me outside and then privately asked if I would help her to enrol in English classes. She wanted to take driving lessons to enhance her independence; however, her poor English proficiency had made this difficult. I assured her that I would explore this. We arranged the first interview on a date she thought her in-laws would not be home.

The next time we met, Tahira looked more comfortable without her in-laws at home. The living room was adorned with religious artefacts such as framed Quranic verses, prayer caps, and prayer mats folded neatly in the corner. Tahira did not disclose Farrukh's diagnosis but recalled that he experienced medical issues since birth.

Farrukh's birth was fine but he was weak, so doctors kept him in hospital. When I was due home, his heart stopped beating so he was put in intensive care. We were both in hospital for two months. When I went home, he stayed for another four months … In his first four years, he was in hospital every month.—Tahira

Tahira recalls that Farrukh's early years were very difficult due to his frequent hospitalisations, requiring a high degree of homecare. Farrukh was wheelchair-bound and non-verbal until he was seven. During this difficult period, Tahira's religious faith helped her cope with her unsupportive family situation, albeit her faith did not lessen the physical labour involved in caring for Farrukh. Tahira recalls considerable challenges in understanding and supporting Farrukh's disability. This was partly due to her poor English proficiency, but also because she assumed that only a family member could interpret her interactions with health and educational professionals. Her father-in-law acted as her translator and so, knew and inevitably influenced the decisions she made. Despite sharing a home, Tahira's in-laws had a detached, almost hostile relationship with her. They frequently excluded Farrukh and Tahira's husband, who also has a learning disability, from extended family gatherings because, according to Tahira, her in-laws were worried that her husband and son would create *"an embarrassing scene in front of relatives"*. However, Tahira, as a first-generation immigrant, relied on her in-laws' support to navigate Britain's education and healthcare systems to provide for her own family. Sometimes this advice was not in Tahira or her family's interests, but her poor English proficiency forced her to rely on their advice.

Like Parveen, Tahira had also experienced child bereavement. Her daughter, Amna had a life-threatening medical condition and passed away when Farrukh was six years old. This was her most difficult period.

> Amna passed away in the same hospital as Farrukh, but different wards so sometimes I was with Amna, sometimes with Farrukh. She was only a baby ... she had brain damage.—Tahira

Unlike Parveen, who benefitted from support from her spouse and her children, Tahira was alone in experiencing these hardships. She had to care for her husband at home and make daily hospital visits to see Farrukh and Amna, with no respite help in the meantime. Amna's death also affected her eldest son, Shehzad who feared he would lose all his siblings to ill health. He became resentful towards Tahira for not giving him time and leaving him with her in-laws. With her youngest daughter Mona, Tahira felt she was slowly growing up on her own and that even when she wanted to support her in school, she could not do so due to her tight schedule.

> Mona really wants me to go to her school. Ever since she joined this school, I haven't been able to attend her parents evenings because of Farrukh. There have been two so far, and both times Farrukh was hospitalized so I couldn't go.—Tahira

In talking to Tahira, I could sense profound sadness that she could not be there for all her children and that slowly and gradually they had stopped confiding in her. When we talk about Farrukh, she felt more confident about her involvement in his schooling. She stated that both Farrukh and herself want Farrukh to be educated ultimately within a mainstream school although she recognises this would be hard to accomplish given the resistance from her father-in-law and from the mainstream schools. Whilst Farrukh attended an infant special school, in the past he was facilitated to attend a mainstream school once a week—this was partial inclusion for Farrukh. The special school at that time felt confident that Farrukh should be able to transition fully into a mainstream school, but upon application, the mainstream school rejected his admission on the basis that Farrukh was not equipped.

> They said he wasn't ready for mainstream, he couldn't speak properly, and if some naughty child hits him and he gets hurt, then you'll blame us. And he wasn't toilet trained either.—Tahira

The visits to mainstream schools altogether stopped after the head teacher stated they could not provide partial inclusion to Farrukh anymore because *"he wasn't ready for mainstream"*. Tahira decided to embrace this as her responsibility, preparing Farrukh to develop that 'specific' skill set that would deem him eligible for mainstream schools. Similar to Parveen's experience of sending Amber to a mainstream, it seemed that the responsibility of ensuring a successful 'inclusion' rested on the mothers rather than with schools. If mothers persisted and showed more interest in mainstream education for their child, then they would be actively be instilled with fear of possible bullying of their child. For mothers like Kiran, and Tahira, the fear of bullying and more importantly the fact their child would not being able to vocalise his experiences of being bullied created a lot of fear and apprehension of mainstream schools. For Tahira, the decision to attend a special school was also heavily influenced by her father-in-law who had sent Tahira's husband to the same special school that Farrukh now currently attended. He was convinced that Farrukh was not 'normal' and that if he were, the mainstream school would have accepted Farrukh by now.

The association of normality with mainstream schools is not new. In my previous research (Rizvi, 2015), South Asian mothers suggested that once their child had joined a special school, they knew that their child would be never be viewed as 'normal'. In talking to Tahira, it became clear that in resisting her father-in-law's decision, she was in fact resisting his imposition of the 'normal/abnormal' label for her child. It may seem harsh and rather judgmental to suggest that Tahira's father-in-law had accepted his own son's fate as deserving a special school placement, which inadvertently also justified Farrukh's placement at a special school based on his own experience of parenting his child. Nevertheless, it is unfair to place the blame of the consequences of ableist structures on the subjects who are most marginalised by those ableist structures. Rajchman (1991) poignantly highlights that the "great complex idea of normality has become the means through which to identify subjects and make them identify themselves in ways that make them governable" (p. 104). Even if they are using the tools and apparatus of an ableist system, they are doing so because that may be the only tool provided to them to navigate their child's disability experience. *Does Farrukh's grandfather require a change in his thinking?* A simplistic answer would be, yes. Parents of children with disabilities who belong to marginalised groups should be responsible for resisting the ableist structures for their children, but they should not be blamed for carrying the greatest burden of navigating an ill-fitting system.

Similar to Saira's experience with Faraz, Tahira suggested that Farrukh was beginning to notice things about his placement at a special school,

> Now he's eleven years old, he can tell the difference between his school and other schools. Sometimes when he visits mainstream school he says to me, 'I want to stay here, I don't want to go home'. I guess because his cousin who's the same age goes to a normal school, so he feels it more.—Tahira

Tahira highlighted that it was becoming increasingly difficult to explain to Farrukh why he was attending a special school when his siblings and cousins all attended a mainstream school. It seemed to Tahira that the goalposts for eligibility to attend mainstream school kept changing and with increasing resistance from her father-in-law, it had become a distant dream. When I asked about her overall experiences of working with special schools, Tahira suggested that it had improved over the years but that initially it was a relationship based on mistrust and miscommunication.

> I didn't feel listened to ... For instance, I'd explained about my difficulty in reading and signing letters, but they still didn't understand and didn't respond with a solution. I think not being fluent in English language makes a difference. I sometimes feel they're unable to understand my point of view or that I haven't explained to them properly.—Tahira

The current negative media attention and government focus on the supposed lack of social cohesion among immigrant communities reinforces a perception that this is due to a lack of English proficiency. However, it would be too simplistic to attribute all the issues that Tahira has had to face merely to language barriers. This negative government discourse is hypocritical because it fails to take into account the funding cuts to interpreter services and ESOL classes during the last decade. According to one report, the funding for ESOL classes experienced a 60 per cent cut, from £203 million in 2010 to £90 million in 2016 (Refugee Action, 2017). Therefore, people like Tahira are often blamed for disengagement with school and other professionals due to their poor English proficiency. However, LAs have not provided any professional interpreters or ESOL support that would facilitate their engagement with services. Tahira was acutely aware that she needed to be fluent in English because this affected all forms of communication with the school. However, unlike Parveen, she did not have a supportive extended family. No one was willing to attend to Farrukh or her husband if Tahira took time out for ESOL classes. In fact, her in-laws insisted that she should involve her father-in-law in every decision because he was fluent in English, and was a more experienced immigrant. This affected the extent to which Tahira could exercise her own agency in professional meetings because her father-in-law had a very different stance on issues. At times, she would have a bundle of letters from school and health services waiting to be signed and looked at, and she would have to wait for weeks before finally getting her father-in-law to explain them to her.

In addition to constant miscommunication between the school and herself, Tahira also sensed that the school did not trust or value her expertise. For instance, the school had recently suggested that Farrukh should bring packed lunches from home as that might help with his weight. When Tahira started preparing fresh chicken curry with *chapattis,* the school came back and informed her to stop sending packed lunches because he does not eat it. When Tahira suggested that at home Farrukh loved Asian food, they stated that she might be exaggerating. Tahira then made videos of Farrukh eating at home, which she then shared with the school. After watching these videos, the school allowed packed lunches of Asian food. Tahira reflected that she initially felt offended that the school distrusted her judgement, and it was important that the school should recognise that she, as Farrukh's mother, understood Farrukh's routines more than they did. Ultimately, she moved on because she felt that the school was advocating for Farrukh's best interests. Between her father-in-law's excessive influence and the school's presumed expertise on supporting Farrukh, she was stuck in a desperate situation.

In our last meeting, Tahira revealed that since our last interview she had been investigated by authorities,

> We were painting our home, there was some kerosene in a plastic bottle ... I was distracted, I didn't look before drinking. I thought it was water. They [professionals] asked me, 'Did you drink it intentionally?' I know I experience difficulty, but that doesn't mean I drank it intentionally.—Tahira

Tahira relayed that she had been unfairly judged by professionals who suspected she had attempted suicide and so doubted her ability to look after her children. Whilst the social workers did later conclude that this incident had been out of character, and therefore did not feel fit to take further action, nonetheless, Tahira felt her trust in professionals had been betrayed. Whilst she admitted that mothering Farrukh and caring for her husband had not been easy, she stressed that mothering Farrukh was not a burden. Of all the mothers I talked to, Tahira's experience is the most extreme example of professional oversight that measures support in terms of dysfunctionality. Although social workers did not consider that Tahira's state of mental health was dangerous for her children, nonetheless, they still left her to look after two disabled people (her husband and Farrukh) without providing her with extra help at home. This poses questions about how the state exercises power and scrutinises mothers within their personal spheres but, as in Tahira's case, offers them very little support.

SHEHNAZ

I obtained Shehnaz's contact details from Anokha, although they acknowledged that they knew little about her background because she was inactive within the group. Shehnaz is a British-born Pakistani and a Sunni Muslim whose family is originally from the Punjab; she is fluent in English and Punjabi; however, interviews were conducted in English. Shehnaz's family have resided in Britain for decades and Shehnaz had lived in Southwest England all her life, where she attained A-levels. She lives with her husband, her five children and her in-laws. Her husband is her cousin and also a British-born Pakistani Sunni Muslim, and works as a taxi driver. Two of her children are Statemented and attend special schools; nine-year-old Amna has global developmental delay and six-year-old Tariq has severe learning difficulties (SLD). Interview schedules were difficult to maintain because Tariq was often sick and so was at home, which made it impossible to visit Shehnaz, resulting in numerous delays. Nonetheless, our final interview, which Shehnaz could have refused, was conducted via telephone at Shehnaz's own suggestion because Tariq's and her own health had worsened.

My first meeting with Shehnaz was held at her home; she lived in semi-detached house which was on the same street as Tahira. We sat in her living room where the oxygen machine was parked prominently but neatly in one corner of the room alongside other medical equipment. "*That's for Tariq*", Shehnaz explained as she caught me looking at the machines.

Amna's older than Tariq, she was born in 2005. It was a normal pregnancy, full term, normal birth, no diagnosis or anything. She was like any other child, normal, bouncing, bubbling and everything. When she started infant school, we had a bit of a crunch time because I think it was the move to a new place, new teacher, new environment, so it took a little while for her to settle. In Year Two, we sort of realised she was falling behind in her school work, she was taking longer to understand things, and she needed a lot more help with self-care.—Shehnaz

Before Tariq was born, Amna had been the youngest in the family and the apple of everyone's eye. When Amna needed a little more help as a toddler in self-care or eating, Shehnaz did not want to make too much of it. To her, all children developed in different ways and at a different pace. Even when the community paediatrician assessed Amna at the age of four and diagnosed her with global developmental delay, Shehnaz believed that she just needed "*extra tender love and care*".

Shehnaz recalls that she involved the school early on in assessing Amna's needs and arranging provisions for her at school. Amna attended Greenwood Primary School[5] at that time, which is a mainstream school. Initially, Amna was on School Action Plus which entailed all her needs being provided for by her school, with the school consulting with the local authority for any professional advice on health, education or social work. For a while, Shehnaz thought that this did help Amna to adjust to her new class setting, but gradually she realised that the responsibility of ensuring that Amna kept up with her class rested with her not the school. Similar to Parveen's mothering experiences with Amber, all of Amna's workshops

[5] All participants and their personal details have been anonymised in this study.

took place during school time on the other side of the town and the school expected Shehnaz to provide the pick and drop service for Amna.

> She went to handwriting workshops, self-care workshops, and care-therapy workshops which sort of help with her balance so that had to be done outside school. They were held on a normal school day and they would be in different centres which meant I had to pick her up from school and take her to a two-hour workshop, continuously for six to eight weeks for each workshop, then take her back to school. This took, you know, a lot of time off her school. She was really emotional.—Shehnaz

Being out of school for up to four hours a day adversely affected both Shehnaz and Amna. Shehnaz stated that she struggled to give equal attention to her other children, especially her son, Tariq who had complex medical needs. Amna found her whole schedule overwhelming, feeling a lack of belongingness to her school. Shehnaz suggested that Amna started to refuse to go back to school after the workshops ended, because she felt increasingly self-conscious that her classmates would notice her coming in late and *"figure out she's got special needs"*. Shehnaz stated that, at times, she pushed Amna to attend school but if she felt Amna was too emotional then she would just take her home. This situation persisted for nearly two years, however, when it became patently clear to Shehnaz that not only were Amna's needs not being met by the mainstream school but that she was not enjoying school either, Shehnaz decided that it would be best for Amna to attend a special school.

> When I spoke to Amna's school and said, 'well if you can't meet her needs what other options does she have?' Again they said it must be a special needs school.—Shehnaz

Amna's mainstream experiences question the purpose of locational inclusion if it excludes disabled children. Notably, Shehnaz did not feel that teachers had excluded her child, rather the whole system and policies were non-conducive for disabled children such as the limited resources within mainstream schools and the demands placed on teachers in a class of 30 children. This inevitably meant that, as a mother, Shehnaz faced difficult choices about leaving the mainstream system altogether in her child's best interests. In a collaborative project by Runswick-Cole (2011) exploring the challenges that children and young people, parents and professionals face within inclusive education, she found that families consistently experienced exclusion within mainstream schools. Despite school-wide inclusive policies, disabled children were unable to access many parts of the school, and were often kept separated from their non-disabled peers (Runswick-Cole, 2011). It seems that Shehnaz and Parveen's experiences resonate with this "*form of apartheid of disabled pupils*" (Runswick-Cole, 2011, p. 116). Even when mainstream schools want to promote inclusiveness, they are penalised by having to compete in school performance league tables as well as facing budget cuts. The National Education Union (2019) conducted a survey of 1026 primary and secondary school teachers working across schools in England, and found the number of SEND support staff is a quarter of what it was in 2009. In another survey by the National Association of Head Teachers (NAHT, 2018), nearly a million disabled children did not receive any additional SEN funding in their schools. Against a backdrop of austerity, inclusion becomes an avoidable expense rather than be viewed as a child's right to a mainstream education. It was interesting to note that as soon as Shehnaz made up her mind about transferring her daughter to a special school, Amna's mainstream school went to great lengths to help with the paperwork and the Statementing process to ensure Amna could get a special school placement. However, since this was an in-year-admission from a mainstream to a special school, the local authority required Shehnaz to apply for a Statement that highlighted Amna's needs and how they were not being addressed at her current primary school. The process of preparing 'evidence' for the Statement required Shehnaz, rather than the school, to complete all the paperwork, coordinate with various professionals, as well as hold meetings with the local authority. Nonetheless, once Shehnaz had crossed that hurdle, Amna was given placement at a special school and was able to resume her studies in Year Four. The special school offered Shehnaz the opportunity to share responsibility for Amna's academic and social inclusion with them, and generally enjoy a more equal relationship. Importantly, all workshop sessions were offered within the school premises which meant that Amna did not have to constantly leave the school premises, and so did not have to play catch-up anymore. Even with the new special school placement, Shehnaz expressed that she had adjusted her expectations with regard to how the school could support Amna,

> As long as their [her child's] needs are met, not 100 percent because each school has their own [SEN specialisation] categories, but if the school's 80

percent meeting my child's needs, and not meeting for 20 percent, then that's good enough. If you look at every detail, you'll never find a school. There's never going to be a perfect school that ticks every single box.—Shehnaz

Shehnaz was willing to accommodate the school provided that it met 80 per cent of her own criteria, trusting that the school would eventually fulfil the remaining 20 per cent gap between her ideal versus her acceptable school support in the long-term as they became more equipped in how they should support her daughter. She adjusted her immediate expectations so that the school was adjudged, in her own words, *"good enough"* for her children even if it was not ideal. Her views also broadly reflect how education increasingly operates in a commoditised free market, with more agency resting with decision-makers (Gorad, 1999). One example of this free-market model is the UK government's initiative for local 'one-stop shops' which would offer advice, support and information for families of disabled children (Children and Families Act, 2014). Despite the freedom that is intended to result from such marketization, mothers like Shehnaz still experienced fewer options which resulted in limited agency to choose. When I asked her why she had chosen a special school rather than expressing a preference for a better and more inclusive mainstream school, she suggested that her positive experiences of working with a special school for her younger son, Tariq had influenced her decision to choose a special school for Amna. Tariq's school was not only well equipped to support children with different abilities, but Shehnaz was also able to manage her time, resources and her relationship with them.

Shehnaz was far more equipped with insider knowledge of the special education system which allowed her to advocate for and embrace a more equal relationship with the school. Her decisions were informed by her own experiences of being educated in British schools and being adept at knowing the roles and responsibilities of different professionals. For instance, she knew that she could check the availability of places at each school for Amna, and if the local authority was delaying the placement without a probable cause,

At that time of year [in-year-admission], my chances were quite slim compared to those parents who'd applied the previous year, but I was quite lucky because a child was leaving Chester Primary Special School when we applied for Amna. As soon as I heard that, I went to the council and asked them why they couldn't give Amna that school place.—Shehnaz

When Shehnaz discussed school placements, she talked about taking up the *"fight"*, *"arguing"* or making a case that her daughter was being failed by the system. She was not willing to settle for a mainstream school if it only provided lip service to the support that should be available for Amna. However, not all people are as equipped to take issue with their local authority, nor are they as fortunate to find a school with an open place for their child. According to one government statistic, more than 4000 disabled children were without a school placement in 2017.[6]

When we discussed Tariq, her experiences of navigating the special education system were very different from her demanding experiences with Amna. It seemed to her that it was relatively less difficult to secure provisions when a child has complex medical needs, rather than when a child has an invisible disability such as global development delay. Tariq was born with a rare genetic condition and from the moment he was born, Shehnaz knew he had serious medical issues.

> Within ten minutes of his [Tariq's] birth, the doctor said he had low blood sugar and then nurses and lots of professionals came in. It was obvious something was wrong because with a normal birth you don't have all these people involved … I was so upset, I had no idea of what was going on. He was taken straight into intensive care and I was just left in this empty room without my child.—Shehnaz

Post-pregnancy, Shehnaz recounts how traumatic it was to learn that her young baby needed an operation. In the first two years of his life, Tariq was hospitalised for a majority of the time and underwent seven major surgeries. His medical condition and resulting hospitalisation led to Shehnaz feeling severely depressed, exhausted, and at one point being hospitalised herself with pneumonia. During this time, she recalls how her husband's emotional and practical support helped her to recover, and was vital to running the household. Unfortunately, this came at the price of financial hardship, because her spouse suddenly became the primary carer for four of their children and had to give up his work so that he could

---

[6] Sally Weale and Niamh McIntyre, 'Thousands of children with special needs excluded from schools', Guardian, theguardian.com, 23 October 2018.

support them at home. For Shehnaz, her own recovery from pneumonia and post-natal depression seemed rushed but she realised that *"the world hadn't stopped"* while she was sick; her other children needed her, and her husband needed to get back to work.

> I thought I have to really look after myself because I don't want to go back to where I was ... you know, my husband did everything and it wasn't fair on him. I think that shook me a little bit then.—Shehnaz

Despite Shehnaz living in close proximity to her mother-in-law and her brother-in-law, her husband took over all the household and caring responsibilities. This was because, as a couple, they were reluctant to involve their extended family network to help due to their previous interference in family matters. Involving relatives meant inviting unnecessary questioning about Tariq and Amna. In talking to Shehnaz, it was evident that she valued her home as a sanctuary for her children; she felt that excessive professional and family interference had already made her children aware that they were different compared to other children. In fact, to enable her to bring Tariq home as well as to reduce their frequent hospital visits, Shehnaz undertook emergency medical and daily care training for Tariq as a way to restore what she termed her *"own normal"*. Shehnaz engaged with Tariq's medical needs for the convenience of her whole family, reducing the extent to which her family's routine was structured around professional schedules thereby taking control of her circumstances and responsibilities.

Unlike with Amna, Shehnaz considered herself fortunate that Tariq had a Statement from the start, and that his placement at a special school provided weekly visits to a mainstream school. There were no unrealistic expectations thrust on Shehnaz with regard to supporting Tariq during school time. His teachers valued Shehnaz's input and expertise and made sure her assessment of Tariq's outcomes were included and addressed in his annual Individual Education Plan (IEP) meetings.

> They're just fantastic. They supported him from day one because it's a special school and they cater for children who have additional or special needs. They all have training in administering his PEG, giving him oxygen.—Shehnaz

Shehnaz reported that her experiences with Amna's mainstream school had been burdensome. They had made sure that she was aware that every step they had taken to accommodate Amna was something outside of their normal routine, as if it was an extraordinary undertaking to provide workshops within the school's premises to help Amna's learning. With Tariq's school, Shehnaz felt immense relief that they proceeded without ever making her feel that they were doing her any favours. Like Parveen, Shehnaz no longer felt that she was being punished for having a child who required extra support. It is interesting to note that both Parveen and Shehnaz did not view special schools as second-class educational settings; they found special schools to be more inclusive and certainly more understanding with regard to partnering with parents to support disabled children.

MARIA

Maria is an active member of Anokha. Before meeting her, we held a telephone conversation so I could learn about her background and responsibilities. Maria is a British-born, second-generation Pakistani Sunni Muslim. She is fluent in English, Punjabi and Urdu. Like most mothers in my research, she requested that interviews be conducted at home during school hours. Interviews were primarily conducted in English, with a few common Punjabi and Urdu phrases interjected into our discourse. Her parents settled in Britain decades ago from Sialkot, one of Pakistan's more affluent cities in the Punjab. Maria was educated in Southwest England, attaining a Bachelor's degree. She married her first cousin who was from Sialkot, and who emigrated to Britain to join Maria after marriage. He works as a taxi driver, whilst Maria is a full-time housewife and primary caregiver for their two disabled children, 14-year-old Aamir and five-month-old Saman, as well as their non-disabled children Azlan, Harris, and Zeenat. Maria's is a single-unit family, although her parents live close-by. Our interviews revealed that Aamir and Saman have a rare degenerative metabolic condition (the name of which I am withholding in order to maintain their anonymity). Aamir uses the same disability service as Tahira's son, Farrukh. Maria identifies as being religious and British, and has not re-visited Pakistan since her wedding.

Maria lives in an ethnically rich neighbourhood of Pakistani and Somali families. The local high street boasted both Pakistani and Somali owned halal butcher shops, barber shops, and ethnic apparel shops. Maria lived on a street of semi-detached houses, with front entrances displaying reflective plastic stickers exhibiting Islamic prayers or phrases such as *Bismillah* (in the Name of God) as Muslim families often do to bless and sanctify their homes. When I first met her, she was holding Saman, her five-month-old daughter in her arms. Her living room had a similar set up to Shehnaz, in that one part of the room contained assorted medical equipment whilst the remaining part of the room contained modest furniture.

> Aamir is my oldest son, he was born full-term with no complications in delivery or during the pregnancy ... it's just that, as he developed, we noticed he was very slow in developing.—Maria

Maria revealed that she knew very early on that Aamir needed specialist medical attention; this became more apparent after Azlan was born a year later and Maria observed that Azlan was developing faster than Aamir. However, Maria's main grievance was Aamir's misdiagnosis, which led to a late correct diagnosis, and her concerns being dismissed by educational and medical professionals during Aamir's first five years.

> Every time we went to the doctor, we'd say he doesn't do this or that, he doesn't respond when we call him. He was my first born and I didn't know any different ... they would just say 'don't worry about it every baby is different, they develop at different speeds so don't worry too much'. Then as he got older, there were more things he wasn't doing and when they checked him they said maybe he's got developmental delay. All this time, it was developmental delay.—Maria

When Aamir was nine months old, Maria learnt that he was hearing impaired. He was born in 2000 at which time universal screening for hearing impairment at birth was not conducted. She recalls her frustration with medical professionals because all his medical conditions ranging from breathing issues, the inability to use his muscles, his difficulty in walking and eating were viewed as symptoms of Aamir's developmental delay. This was a difficult period in her life because Maria was not satisfied with the medical explanation. At home, Maria could see that Aamir's health was deteriorating more rapidly as the days passed, becoming ever more dependent on her. Finally, when Aamir turned six, and with the help of his consultant, Maria was able to push for Aamir to undergo a thorough medical examination.

When he [Aamir] was six, they did blood tests, he had lumber puncture, muscle biopsy ... he was diagnosed with XYZ.[7] It is a metabolic and degenerative condition. I just cried my eyes out ... it took a couple of days for the news to sink in. They told me Aamir wouldn't live beyond the age of eight and he was six when he was diagnosed. I thought 'oh my God, he's only got two years left'. Then it sort of sank in.—Maria

Maria felt a lot of resentment towards medical professionals for wasting time and taking away the precious years that she had left with Aamir. Suddenly, after her realisation that "*he's only got two years left*", her goal was no longer fighting for Aamir's needs to be accurately identified, but rather delaying the imminence of his death. She requested second medical opinions but unfortunately found that repeat tests yielded the same results. Like Parveen, Maria began to conduct her own research, utilising medical websites to gain a better understanding of Aamir's condition. She recalls that she had not observed any of the symptoms which were characteristic of his condition at the time of his diagnosis, nonetheless, as time went by, she began to see Aamir displaying all the symptoms that doctors had cautioned her about. Notably, again like Parveen, Maria also believed that matters of life and death were not dependent on a doctor's diagnosis; rather that only God ultimately knew Aamir's fate. Armed with her well-researched medical knowledge about Aamir's condition and her religious faith, she gained the confidence to navigate this difficult period. As was the case with Parveen's daughters, Sehr and Amber, Aamir's condition is so rare that there are only about a hundred known individuals in the world with this diagnosis; this meant that the intervention itself was still in its trial phase, and that there was very limited information about this condition available to the public.

Akin to Saira and Kiran, Maria was also critically aware of the problematic construction of normality/abnormality.

I'd say my responsibility towards Aamir hasn't decreased. That responsibility will always be there regardless of my other children because, I don't know how to say it, they're normal.—Maria

---

[7] I am not stating the precise condition because Maria's son, Aamir has a very rare disease which could easily identify my participant and her son, thereby breaching their confidentiality.

It is essential to understand that the construction of normality/abnormality is mutually constitutive of the existing experiences of carer roles, and of the simultaneous mothering of children in the family who have a disability and children who do not; it is also constitutive of other categories such as culture, religion and gender. Maria highlighted that while she obviously loved all her children equally, the level of care and responsibility she exercises towards Aamir and Saman is far higher than the care and responsibility towards her "*normal*" non-disabled children. Maria's construction of normality/abnormality differed from Saira and Kiran because her son has a visible disability, and she was not constantly having to convince other people that her child has a disability. However, like Saira and Kiran, Maria used language that she had been socialised into and that viewed her son as 'not normal'. Even when Maria highlighted that she refuses to see Aamir merely as a medical condition, she was constantly pulled back into an ableist medicalised world of doctors who were excessively interested in learning more about Aamir's rare medical condition; their probing medicalised gaze reminded her that Aamir is not normal, and that she should see him as such.

Maria also expressed an overwhelming frustration with the negative perceptions of consanguinity and its link to Aamir's diagnosis. Many mothers reported that one of the first questions they were asked by medical and other professionals was whether they had a cousin marriage.

> He's [Aamir] got this condition … [I] had a first cousin marriage which I understand could've caused not just this condition but any condition … I don't see him just as a condition that he's labelled with, but we do as best as we can.—Maria

Interestingly, Maria's interview revealed her sense of 'double consciousness'; she acknowledged that many professionals viewed the consanguineous nature of her marriage as the cause of her child's disability. Indeed, she also recognised that children from cousin marriages also had a greater chance of inheriting their parents' other genetic characteristics such as a tendency towards heart disease, diabetes or cancer. Existing health studies have found that male gatekeepers within minoritised families often restrict information about the inherent risks related to cousin marriages, and that particularly women within the family are not made aware of the perils (Darr et al., 2013). Significantly, mothers in my research like Parveen, Maria and Tahira who discussed cousin marriages were not only aware of the inherent risks, but also expressed how difficult the decision had been to have children and how it affected their future family planning. It was not simply a case of knowing the risks and then making a medically informed decision; Maria had to decide whether to have another child knowing they already have one disabled child. Maria revealed that she not only received genetic counselling but was also asked by medical professionals to consider an abortion if tests confirmed that her foetus had a genetic condition.

> [Medical professionals] said if tests confirm the condition … you've got to be prepared to abort the pregnancy. I said, 'I won't do that, we're going to accept her regardless of whether she has a condition or not'. They said it was my choice.—Maria

Maria's reproductive practices became the site of conflicting interests between her own views regarding mothering, and medical experts who advised that she should "*be prepared to abort the pregnancy*" if the foetus was not healthy. Whilst she was aware that her decision to continue with her pregnancy despite knowing that her baby might be disabled may be viewed as careless or even selfish, nonetheless, she argued that she had unconditional love for, and was prepared to do everything to support her child. Her stance was possibly an attempt to arrive at her own understanding of what mothering a child entailed, and that disability is merely one aspect of her child rather than the defining marker. In my conversations with her, she critically engaged with her own choices as a woman who was free to exercise her reproductive rights, and her stance on having a disabled child. Maria's feminist and disability perspectives are not at odds with each other. However, it is difficult to ascertain whether her religious positioning bridged these two opposing perspectives, influencing her decision to keep her baby since Islam strictly prohibits abortion unless the pregnancy is detrimental to the mother's life. Nonetheless, Maria's experiences exemplify how complex such decisions can be in the context of cousin marriages, and how these decisions may be perceived by professionals.

The correct identification of Aamir's disability did eventually help put the right provisions in place, giving Maria a better sense of which placements to consider for him. Due to Aamir's complex medical needs, his network of professionals did not encourage a mainstream school placement. Maria was also concerned about the different priorities at mainstream schools.

> Some schools don't want their [academic attainment] levels to drop ... if children lag behind, they won't allow those children to bring it down.—Maria

Maria suggested that the mainstream curriculum was incompatible for her child, and the inclusion checklist such as railings, ramps and access to playgrounds only existed on paper but not in practice. Moreover, she felt that mainstream schools were not fully invested in disabled pupils because such children would lower their academic performance; indeed, she suggested that this mindset was endemic within Britain's education system. A small-scale study into the school admissions process in Britain echoes her concerns (Office of the Children's Commissioner, 2014); parents of disabled children reported that some mainstream school staff actively discouraged them from applying to their schools, or implied that their child would progress more at another school. In some cases, this advice was offered based on a lack of parental awareness of the admissions process, or by making overtly emotional appeals to parents (Office of the Children's Commissioner, 2014), as Kiran and Tahira also reported.

Interestingly, Maria never assumed that Aamir would thrive in every special school setting.

> We visited Cavendish and Brindley [special schools]. At Cavendish I came out crying, I just couldn't send Aamir there. Most of the children were in wheelchairs. Aamir had just started using his wheelchair but wasn't full-time, so I was thinking, 'how do we send him to a setting where it's normal for everybody to use wheelchairs?' ... I wasn't happy.—Maria

Mothers in my earlier study (Rizvi, 2015) also discussed the negative visual imagery of disability when their child first entered special school, which they perceived as their child's segregation from 'normal' society. However, the perceptions of those mothers in my earlier study were unaffected by the physical aspects of special schools (Rizvi, 2015). The visual imagery of disability as portrayed in the media, in public spaces, children's literature, and representations by charities have been explored within existing studies (Garland-Thomson, 2002; Darke, 2004). Interestingly, according to a study by Bagley et al. (2001), parents of non-disabled children also consider a school's physical environment during placement decision-making; however, unfortunately, research specifically examining whether physical environmental aspects within special schools affect placement preferences is scarce. Nonetheless, mothers within my current research like Shehnaz, Kiran and Maria specifically chose special schools which did not resemble a hospital. Maria suggests that the closer her impression of a special school was to her own sense of normality and that did not project feelings of "*hopelessness and sickness*", the more likely it was that she would prefer that setting. Maria also suggested that perhaps special schools should consider rethinking certain aspects of their physical environment, so that they "*appear more inclusive and welcoming to children and parents*" alike.

After visiting and exhaustively comparing various special schools, Maria eventually chose Brindley Special School; however, she still faced a lot of pressure from Aamir's new special primary school. For instance, the school insisted that Aamir should have a PEG fitted in his stomach because of concerns over how long it was taking to feed him orally, and the cost of allocating a dedicated help to sit with and feed him. This echoes the wider socialisation of disabled people and their families into ableist practices from an early stage. School policies and procedures that impose restrictions on disabled children and their families, such as pressuring or coercing parents into approving their children to be fitted with a PEG, are not only physically and emotionally taxing for children and their families but they also reinforce values of intolerance towards children who are not able-bodied. Ryan (2018), in her article, reports on the experiences of disabled women in Britain who were forced to undergo surgery to have catheters fitted solely to negate the need for (and the resulting financial cost of) making public toilets accessible for disabled people. The doctors in these cases stated that these surgeries were correcting 'social incontinence', since these women did not suffer from physical incontinence or a lack of bladder control but rather were simply unable to use public toilets due to accessibility issues. In Aamir's case, Maria knew that his prognosis stated that he would eventually lose his ability to swallow, however, she was not ready to approve a PEG whilst he could still swallow and enjoy the taste of food. I asked Maria whether it would have made life easier for her at home to have the PEG fitted,

> I didn't get it done because I wanted that independence for him as long as he's physically capable ... that's one thing he shares with us, sitting at the

dinner table and eating with us as a family. I'm not going to take that away from him just because they [school] felt it's too time consuming, or they've got other children who they need to get to … but they put on pressure.—Maria

Maria recounted receiving appointment letters from consultants to discuss the possibility of a PEG for her son, as Aamir's school eventually involved medical professionals. She insisted that doctors test Aamir to check if he could still swallow, which proved affirmative, and which allowed her some bargaining power with doctors and educational practitioners who could not proceed without her consent. Unfortunately, a few months later, Aamir caught a chest infection which led to his health deteriorating rapidly and adversely affecting his appetite. During this time, Maria reconsidered the issue of the PEG because she was concerned that Aamir may stop eating altogether. She approached the consultants herself and agreed to have a PEG fitted as a contingency plan for those times when Aamir was too ill to eat orally. This was a huge decision for her, but she was satisfied that it had not been rushed, and that at home at least, Aamir would still be able to enjoy his independence and the experience of "*sitting at the dinner table and eating with us as a family*".

In contrast to her experiences with Aamir's primary schools, Maria had a positive experience with Aamir's secondary school. The special primary school was a feeder for the special secondary school, and shared the same headteacher. Maria described the secondary school site as being bigger and more suitable for Aamir, and described her relationships with teachers as more equitable. It is difficult to ascertain whether this more equitable relationship was achieved as a consequence of the difficult decisions that she made during Aamir's primary school years, such as the decision to fit Aamir with a PEG, which the secondary school would have benefitted from. Nonetheless, from our conversations it was clear that Maria felt that Aamir's secondary school valued her expertise in supporting Aamir.

They asked what kind of routine I use at home, what things they could adapt from me … they wanted to know if there's anything I'm doing at home that they can copy … they're where I'm coming from.—Maria

Maria's relationship with Aamir's secondary school exemplifies how collaboration can take various forms, and is similar to the experiences of working-class Mexican parents in the study by Moll and Greenberg (1990)

on 'Funds of Knowledge' in the USA. They examined household learning practices and values which schools could emulate in order to improve children's learning outcomes. The aim of the study was to work alongside minority communities, utilising their family capital to building trusting relationships. Although Moll and Greenberg (1990) researched with non-disabled children, nonetheless, their findings are mirrored in how Maria's own household practices which supported Aamir were welcomed by his school. Maria revealed that teachers, medical professionals, the social worker and Maria herself all respected each other's perspectives, and knew why they were taking such a stance within the relationship. This is possibly because special schools are more flexible compared to mainstream schools and more accustomed to accepting suggestions from other professionals and from families as part of their IEP exercise, encouraging parents to be active in their child's educational goals. However, it is also possible that Maria's own insider status to British schools provided her with the necessary knowledge and confidence to demand a place at professional meetings, even successfully negotiating the conditions that she considered would benefit Aamir. Consequently, Maria displayed greater trust in Aamir's school, reporting fewer problems within home-school communication.

ALINA

I contacted Alina through snowballing. Parveen, another participant in my research knew Alina and upon my request asked whether she was interested in participating in my research. Alina is a first-generation Pakistani immigrant and Sunni Muslim from Islamabad, the capital of Pakistan. Alina and her husband are both in the UK on work visas but want to settle here permanently; Alina is a full-time doctor in a nearby NHS hospital and her husband is an IT consultant. Alina only joined her husband in the UK five years ago after her only child, eight-year-old Imran was diagnosed with autism in Pakistan, where Alina knew his needs would not be supported. She speaks Urdu and English at home and lives in a single unit family.

Alina lives in a leafy suburb on the outskirts of the city. Her house is located ten minutes from the motorway which is ideal because her work as an NHS doctor is in a neighbouring city. I was fortunate to arrange an interview with Alina because she had irregular work hours and was often on-call. Our first meeting was at her home, but subsequent interviews were held at a local children's play centre which her son, Imran enjoyed visiting. In our first meeting, Alina shared her experiences of living in Islamabad, Pakistan's capital city before she moved to Britain, how she was just beginning to thrive in her career, and why she decided to move permanently to the UK.

> Imran was born in 2006 ... the first time someone suggested he may have autism, I was working in a children's hospital as a paediatric doctor. When I researched it, I found out it was a lifelong disability. I thought perhaps he doesn't have it, perhaps he'll get better. Initially, I was in that phase but quickly came back to reality, I felt that now I've got the diagnosis I should work out what to do.—Alina

Being a medical professional herself granted Alina easy access to an established network of specialist doctors, and gave her the confidence to pursue medical consultations for Imran in Pakistan. However, the fact that she had failed to make sense of Imran's needs earlier despite her medical background did come as a shock to her. A feature that became evident within each mother's narrative was the disproportionate labour of care involved in their lives. Even those mothers who equitably shared the physical aspects of their childcare responsibilities with their spouses still shouldered most of the emotional, mental, and physical labour of setting up their child's support, booking appointments, filling out applications, and meeting professionals. As a result, they did not have the time to process or attend to their own emotional needs because at least one parent, as Alina put it, had to get "*back to reality*". It is also essential to recognise that this disproportionate labour of care cuts across social markers such as the socioeconomic class, immigration trajectory and educational background of these mothers. This may be partly due to the intrinsic value we assign to mothers as inherently life-affirming, emotionally resilient parents in comparison with fathers; indeed, society as a whole views the mother/child relationship as naturally special due to the biological experience of childbirth (bell hooks, 2000). This practice of socialisation romanticises mothering thereby minimising the struggles and the emotional toll it takes on women which only benefits a male supremacist ideology. In this research, this was further complicated by the presence of South Asian cultural values and interpretations of Islamic theology on child rearing. Therefore, notwithstanding the widely differing social positionings of, for instance, Alina as a doctor from Tahira as a housewife, nonetheless, they were similarly responsible for performing the labour of both parents.

During our conversations, Alina also felt it was necessary to clarify that there was no history of disability in either her or her husband's family, that her marriage was not consanguineous, and as such she could not have anticipated or caused her child's disability. The question of consanguinity is often asked of mothers, not only by professionals but also by friends and relatives; a sad result of this mother blaming is that oftentimes mothers choose to address consanguinity head on rather than be blamed for their child's disability.

When Imran was diagnosed, he and Alina were living in Islamabad whereas Alina's husband was living and working in the UK as an IT consultant. After Imran's diagnosis, Alina realised how inadequate childcare support was in Pakistan in general. Every childcare setting she approached was unwilling to enrol Imran because they required him to be toilet-trained and stated that they could not address his needs. Indeed, many childcare settings implied that Imran was not happy, and was reserved around other toddlers because he was not being looked after at home. This was the last straw for Alina, who decided to home-school Imran thereafter, believing that no childcare setting had an accurate understanding of Imran's needs. This decision was particularly difficult for her because she was just starting to gain professional recognition as a doctor at a local hospital in Islamabad. She quickly started researching and educating herself on the educational provisions for children with autism available in the USA, Canada and the UK.

> I was sad I couldn't go to Canada or USA. They have far better provisions than Pakistan and better than UK even, but I couldn't go so I ended up in the UK.—Alina

Alina's immigration trajectory differs to the other mothers in my research. As a first-generation immigrant, her reason for migrating was not due to a transnational marriage or socioeconomic reasons; indeed, her standard of living was far higher when she was living in Islamabad. She migrated to Britain because she recognised that Imran could access better provisions in this country compared to Pakistan, although she suggested not at the benchmark available in North America. Notably, the fact that her husband was already resident here was not a significant factor for her. She suggested that she ultimately *"ended up in the UK"* because her medical degree was not recognised in either Canada or the USA, so she would have had to sit for medical licencing examinations in order to qualify to work in those countries; this would have entailed spending more time studying, leaving her less time to care for Imran. Since her existing qualifications allowed her to start working immediately in the UK and her medical expertise was in demand, she applied as a spouse on her husband's UK work visa and arrived in Britain in 2010.

Another issue that was apparent throughout my conversations with Alina was her sense of detachment from, and resentment towards her spouse. Even though they were living together as a family in the UK, Alina and Imran seemingly lived a detached existence from Imran's father. She revealed that not only was her husband "*in denial about Imran's autism*", but that he also avoided spending time with him. Spousal denial of their child's disability was a common thread within Kiran, Maria, and Alina's narratives although Maria and Kiran suggested that their husbands were still hands-on in terms of their childcare responsibilities and generally very understanding towards them. However, Alina suggested that her husband had no interest in being involved with Imran's caring responsibilities or his educational experiences. Ultimately, everything was Alina's responsibility which affected her own ability to care for Imran.

> I don't get a break, I'm getting more tired day-by-day, I'm just burning out … I'm on my own. I don't have family here to help me and I hardly get any support from my husband. I use carers whenever I'm at work and when I come back, they leave immediately. I pay them myself because I don't have recourse to public funds at the moment. Maybe once I get a UK passport, I can apply for some.—Alina

Alina's struggle to gain any spousal support was all the more pressing because her immigration status did not allow her to apply for any formal support such as Disability Living Allowance (DLA) or respite services. Using her salary as a full-time doctor and some financial input from her husband, she was able to pay for various services such as a carer or verbal behavioural analysis (VBA) sessions at home. However, providing what she deemed as essential services for her son came at a steep social cost. Alina revealed that her busy work schedule rarely allowed her to attend school meetings, and that most of them were either rearranged or had to be conducted over the phone. She was aware that Imran's school viewed her as a passive parent, but this was a price she had was willing to pay to ensure Imran was adequately supported at home. She suggested that the current hardship was worth the future promise of support and provisions that would help Imran gain some independence. She also recognised that schools and local authorities were cash-strapped and were gradually decreasing their level of funding for formal provisions, and that by the time she would be eligible that many of these services may not be available for Imran; nonetheless, Alina was still convinced that Imran was better off in Britain than in Pakistan. She valued Imran's school and considered it significantly superior to the educational settings available in Pakistan; she felt at ease in following the school's lead in educational decisions, having faith that the British special education system had Imran's best interests at heart. Her relationships with Imran's schools symbolise the educational priorities that first-generation immigrants hold; they are more concerned with the inclusion of their child within any educational setting as opposed to pushing for their child to be educated in a particular setting, specifically a mainstream setting. Constantly rationalising Imran's educational experiences in the UK against what he would have experienced in Pakistan had she stayed, mediated her day-to-day decisions,

> I think I'm very compliant ... there was one incident where I felt what they'd [teacher] said was strange, it was Imran's previous school. They wrote me a note saying he'd injured a child seriously with his scratching, I mean they did a risk assessment. What was traumatic for me was that Imran was being kept isolated from his class. Look, I've experienced his hitting first-hand, so I know he's strong and could hurt another person. I mean I didn't like their solution, but I felt if I told them not to treat him this way it might cause more problems for him.—Alina

The fact that Imran's primary special school had decided to isolate him from his classmates as a risk assessment measure deeply disturbed Alina. Whilst she had experienced Imran's physical outbursts herself, she felt that isolation was not a helpful intervention. When she reached out to the school, they explained their rationale for their stance, but to Alina's mind Imran had little understanding of why he was being isolated; more importantly, the school had not addressed the environmental triggers that had potentially led to his physical outburst against the other child. Alina also recognised that if she did interfere or complain about the school's intervention too robustly, they may consider alternative strategies such as temporary term-time exclusions which would severely upset Imran and disrupt his school routine. Whilst Imran's experiences at school improved after he transitioned to another special school within the same academy, Alina felt that she still occupied the status of an outsider with the school. Although she did not deny that the school had helped and supported Imran, nonetheless, bureaucratic inefficiencies frequently caused unnecessary delays. Similar experiences have been noted in Bacon and Causton-Theoharis (2013), a USA-based study with 17 families of children with disability, who also found that equitable educational participation was prevented for some families as a result of school bureaucracy. These bureaucratic delays occurred in situations where educational professionals maintained a position as the principal knowledge bearers, even if they simultaneously displayed genuine understanding of a child's needs or empathy towards parents. With Alina, whilst she understood that services like occupational therapy (OT) and speech and language therapy (SLT) were routed through Imran's school, she suggested that teachers were reluctant to share their resources such as OT videos for home use and insisted that she attend school to learn from them directly. Moreover, every support provided by the school had to first be self-initiated by Alina herself. Such professional laxity may prevent some mothers who have multiple caring responsibilities from developing the necessary skills for their child's needs at home.

Alina also expressed doubts and confusion regarding what her expectations should be from Imran's current special school.

> I find it difficult that I can't make a decision about who to talk to … I don't know if the school is responsible to give out all this information to parents.—Alina

This confusion was further complicated by ambiguity around which school provisions were available to her and which had been restricted by the Home Office due to her immigration status; Alina constantly enquired about all forms of support under the new Code of Practice (CoP) that Imran could benefit form. Alina did not have an extended family or close family friends in the UK to give her an insider's view of how the special education system worked with regard to schools and local authorities. Despite her higher educational qualifications, Alina had to work harder than Tahira to familiarise herself with Britain's special education system. Tahira had access to the relevant cultural capital through her albeit overbearing father-in-law. This illustrates how getting the system to work in favour of your child is not solely a function of one's education and wealth, but also of one's access to knowledge regarding how to navigate the special education system.

In similarity with Kiran's experiences of navigating her son's invisible disability in public spaces, Alina also frequently felt people glaring or passing comments at her when she was out and about with Imran. However, in contrast to Kiran, Alina was acutely concerned that Imran's sudden public outbursts may be misconstrued as characteristic of the behaviour of South Asian families in general.

> Seeing an Asian family, some White people assume we just don't have manners and that kind of impression, so you say sorry even if it's not your fault.—Alina

Perhaps Alina, as a first-generation immigrant, felt this societal judgement and gaze more severely in these moments than Kiran; she was not just advocating for Imran's inclusion as a child with disability but as a first-generation immigrant child with disability who was perceived to be acting in an uncivilised, non-Western way in a predominantly White space. Her reactions to these episodes reminded her that she "*still didn't feel fully at home yet*". However, even in these instances, Alina constantly compared her experiences in Britain with her previous life in Pakistan, and was convinced that the exclusionary and ableist attitudes that she experienced in public spaces within Britain were less pervasive compared to the exclusion that Imran had faced within nursery settings, social gatherings and in general public spaces in Pakistan.

Maham

Maham was also contacted through snowballing. I had asked Alina if she knew any mothers and she suggested her neighbour, Maham may possibly be willing to participate. Maham is a British Pakistani Sunni Muslim. She is a full-time housewife with three children; her eldest child, seven-year-old Daniel has global developmental delay with autistic features. Maham migrated to the UK eight years ago after marrying her British-born husband who is related to her and works as an engineer. Maham lives in a joint family with her husband, parents-in-law and sister-in-law. She is originally from Faisalabad in Punjab, and holds a Master's degree in English Literature; she speaks Punjabi, Urdu and English at home. She suggested her in-laws were greatly involved in Daniel's care and daily schooling experiences. Her mother-in-law and sister-in-law were present during all the interviews which created some problems, but Maham assured me that she wanted them present. I wondered later if she had said this in order to put her in-laws at ease in front of an outsider. In this narrative, I will for the most part focus on Maham's voice and not reference the testimony of her mother-in-law and sister-in-law, except in those moments when her testimony becomes inseparable from theirs. These moments sometimes occur when Maham relies on her in-laws to recall a memory, or when she is actively referring to their testimony.

Maham lives just a few streets away from Alina. She lives with her husband and three children next door to her parents-in-law and her sister-in-law in two adjacent semi-detached houses. I was warmly welcomed into her reception room, where her mother-in-law and sister-in-law were waiting. Maham immediately began talking about Daniel.

> He was full-term ... he showed some signs early on, but we didn't pick them up because he was our first child. We thought all children are like this because we didn't really understand it at that time.—Maham

Daniel was born in 2007 and was the first child in Maham's extended family. Living in a close-knit joint family meant that Daniel received love and attention from his aunts, uncles and grandparents. Maham recalls that from as early as seven months, she and other family members started to notice that Daniel was not feeding properly and was unable to maintain eye contact with anyone. At around that time, Maham's sister-in-law sought the health visitor's help to request further medical referrals. Before Daniel had turned three, he was diagnosed with global developmental delay with autistic features. By this time, Daniel was attending a mainstream nursery with 20 other children in his class, and whilst the nursery did help in supporting his Statement, their lack of one-to-one support and rather overcrowded classroom meant that Daniel often refused to go into the nursery. Maham realised that mainstream settings were overwhelming for Daniel; he was often isolated in a separate room with a TA when teachers felt they had to focus on the other children. Maham felt that Daniel had no structure or specific programme that was designed to keep him engaged, left to play alone for most of the time whilst the other children were learning in the classroom.

> I wouldn't call them bad experiences, it's just that they didn't understand him because they were normal teachers who work with normal children ... they didn't understand Daniel has special needs that are different to other

children … it's about having understanding, everything comes from understanding.—Maham

Maham was reluctant to criticise the mainstream nursery because they had helped her apply for a Statement for Daniel. Nonetheless, she admitted that the nursery did not have a specific plan on how to support Daniel's educational and social development because "*they didn't understand Daniel has special needs*", and only seemed to focus on working with children without any disabilities. Maham's views also resonate with other parents of children with disability who may seek special school placement after disappointing mainstream experiences. For instance, parents in the study by Runswick-Cole (2007) who appealed for special school placement emphasised the lack of a tailored curriculum and less provisions within mainstream schools, as well as their child's exclusionary experiences within mainstream school. Runswick-Cole (2007) also postulated that parents who prefer special school settings from the outset often trust professional expertise. This trust may not be because special schools have better trained teachers per se, but rather because of the hesitance of mainstream schools to support disabled children. Like Maria and Kiran, Maham also recognised that Daniel's educational experiences fell outside the domain of "*normal teachers who work with normal children*". She revealed that Daniel could not do all the classroom activities in his nursery setting that other children were doing easily. For her, mainstream settings only suited a certain group of children, and Daniel was not a part of that group. However, her perceived view on the failure of mainstream schools to be inclusive ultimately reflects a wider failing of an education system that places more importance on the education of non-disabled children over the education of disabled children.

Maham, her husband and sister-in-law researched special schools in their area so that when Daniel started formal schooling, he would be in a special school. However, their attempts to find a special school placement for Daniel were less successful than Shehnaz's; they were informed by the local authority that there were no available places in any special school in their area for Daniel during the school placement process. The local authority gave Maham the choice of home-schooling Daniel for a year until a special school place became available, or she could continue sending him to the existing mainstream nursery which would receive extra funding from the local authority to support Daniel. Both options provided by the local authority were not ideal for Daniel and Maham. She and her family desperately wanted Daniel to attend a special school and felt that he should not be sitting at home when other children were in school.

Maham: "We were looking for a special needs school, weren't we? ... even though the nursery offered to keep him there."
Sister-in-law: "He ended up staying at the nursery for an extra year, the reason being he couldn't get a [special school] place ... he had issues there because they weren't proper special needs teachers, they didn't understand what he needed. So that one year did go to waste, it put him back a bit."

Unfortunately, Daniel's experience is not an anomaly; at the start of 2018, more than 2000 children with special needs were waiting to be allocated a place within mainstream or special schools (Weale & McIntyre, 2018). Despite being a first-generation immigrant herself, her husband's close family circle afforded Maham an insider's perspective and knowledge into the workings of the British education and special education systems. What is striking is that despite proactively conducting extensive research on the various special schools available locally and applying through the formal channels, Daniel still ended up losing a whole year of formal school education due to the non-availability of places. Stating a placement preference does not mean parents are offered their ideal placements by local authorities; even when the goal of 'greater choice and control for parents' is embedded in government policy, the reality of the shortage of school places and cuts to existing school budgets means that parents like Maham and Shehnaz are nonetheless dependent on local authorities to make the final decision and award a school place (Bajwa-Patel & Devecchi, 2014; DfE and DfH, 2015).

Maternal placement preferences are also mediated to some extent by an individual's immigrant trajectory. As first-generation immigrants, Maham, Tahira, Parveen and Alina were still in the process of understanding how the British education, social and healthcare systems worked.

I should tell you, at the start I didn't know what special schools were like in the UK. I hadn't observed them or had experience of them.—Maham

Maham's own personal experiences of being schooled in Pakistan meant that it took her time to become familiar with schools' expectations of parental roles, and she should become involved in Daniel's education. She only gained experience and knowledge about how UK schools operated after Daniel eventually started school, and she started engaging and working with school staff. Reay's (2004) analysis of how different aspects of cultural capital affect how mothers participate in their children's education captures this very poignantly,

> Cultural capital is implicated in mothers' ability to draw on a range of strategies in support their children's schooling. As well as financial resources, key aspects of cultural capital such as confidence in relation to the educational system, educational knowledge and information about schooling all had a bearing on the extent to which mothers felt empowered to intervene in their child's educational trajectory and the confidence with which they embarked on such action. (Reay, 2004, p. 78)

Obviously, this is further complicated by the two-tiered education system in Britain that promotes a distinction between children with disability and non-disabled children. Therefore, possessing knowledge about the inner workings of mainstream schooling does not guarantee that one would also have the knowledge and competency to successfully engage with special schools and the wider special education system.

Eventually, a year after her initial school placement application, the local authority informed Maham that her second preference school had finally got an available place and Daniel could be enrolled straightaway. During the following two years, Maham felt that his special school was significantly better at providing formal support to Daniel than his mainstream nursery had been, nonetheless, she also observed that the school were short of funds which had left them with inadequate staffing and resources. She decided to raise her concerns with the school; however, they suggested their class sizes were smaller than other local special schools.

> His school said there is more one-to-one interaction with pupils here than in other special schools, and that other special schools are more crowded. So that has silenced me a bit because I know Daniel needs one-to-one. But I do want to find out about other schools.—Maham

Daniel's special school staff also cautioned Maham about the risk of transferring her son to another special school, only to find out that they suffered from more severe funding deficits. Although Maham was still researching for better schooling options for Daniel, she felt in a bind; if she did switch Daniel to another school, there was no guarantee that it would be better funded than his current school, however, if she never switched then she would never know if there were more appropriate provisions available for Daniel. What became increasingly evident from my conversations with mothers in my research in general, and Maham in particular, was that their placement preferences were primarily driven by a desire to ensure that their child would receive the best possible provisions that would facilitate their learning, before they could actually focus on their child's learning. Whether they had expressed a preference for a mainstream or a special school, mothers were fighting for inclusion for classroom learning and adequate provisions that would ensure their child felt socially included within their education setting.

As Alina and Kiran had done, Maham also shared her own experiences of navigating her son's invisible disability within public spaces.

> He [Daniel] has a habit of putting his hand in other people's pockets to grab their mobile, and obviously they glare at him and tell him off ... he looks normal and not everyone's used to this behaviour. Of course, we say sorry and hurry away ... we don't go into detail as much.—Maham

Maham revealed that her trips in public spaces were short in duration due to how Daniel's behaviour might be perceived in public. She was always accompanied by her in-laws to help care for Daniel which also relieved her from facing public hostility alone. In talking to Alina, Saira, Kiran and Maham, I wondered whether gender was a factor in influencing how unruly their child's behaviour was perceived in public spaces. Boys with an invisible disability such as behavioural needs often attract greater public hostility than girls, especially boys from ethnic communities who may be perceived as a threat (Laura, 2014).

In my conversations with Maham, we also touched upon her religions worldview and how it shaped her mothering responsibilities in the present and in the future.

Maham: "My children are my priority. Once I'm free of my responsibility towards my [non-disabled] daughters, I can devote my life to Daniel. That's my thinking for the future. I've already lived my social life ... I'm answerable to God, aren't I? God made me a mother and a

	huge responsibility comes with that. If I abandon him to enjoy my social life, that just wouldn't be moral."
Mother-in-law:	"But it's not just about answering to God dear. I mean, he's part of you, you could never leave your flesh and blood."

As a mother and a practicing Muslim, Maham viewed that her religious salvation would be ultimately through mothering Daniel. Islamic theology teaches that all parents are accountable for how they care for their children, a parental caring responsibility that extends to parents of children with disabilities and non-disabled children alike (Hamdan, 2009). Existing research with Muslim families into their religious explanations of disability, has considered this religious interpretation of parenting as unique only to a Muslim context (Croot et al., 2008). However, several studies conducted with families belonging to various religions found that their participants also ascribed to such a religious worldview (Poston & Turnbull, 2004). Maham suggested that it was her religious duty to care for Daniel as best as she could, and to refuse would be immoral. Maham's self-imposed binary highlights how religion influenced her sense of morality with regard to what being a good mother entailed. More importantly, her sense of morality and her maternal obligations were seemingly tied to her mother-in-law's expectations of her. Whilst her mother-in-law explicitly rejected Maham's view of morality through a religious lens, nonetheless, she still appealed to her moral compass by insisting that, as a mother, Maham "*could never leave* ... [her] *flesh and blood*" in an assisted living facility.

References

Ammerman, N. (2007). *Everyday religion: Observing modern religious lives.* Oxford University Press.

Bacon, J. K., & Causton-Theoharis, J. (2013). 'It should be teamwork': A critical investigation of school practices and parent advocacy in special education. *International Journal of Inclusive Education, 17*(7), 682–699.

Bagley, C., Woods, P. A., & Woods, G. (2001). Implementation of school choice policy: Interpretation and response by parents of students with special educational needs. *British Educational Research Journal, 27*(3), 287–311.

Bajwa-Patel, M., & Devecchi, C. (2014). 'Nowhere that fits': The dilemmas of school choice for parents of children with Statements of special educational needs (SEN) in England. *Support for Learning, 29*, 117–135.

Bell, E., Andrew, G., Di Pietro, N., Chudley, A. E., Reynolds, J. N., & Racine, E. (2016). It's a shame! Stigma against fetal alcohol spectrum disorder: Examining the ethical implications for public health practices and policies. *Public Health Ethics, 9*(1), 65–77.

bell hooks. (2000). *Feminist theory: From margin to center*. Pluto Press.

Bhatti, G. (1999). *Asian children at home and at school: An ethnographic study*. Routledge.

Bywaters, P., Ali, Z., Fazil, Q., Wallace, L. M., & Singh, G. (2003). Attitudes towards disability amongst Pakistani and Bangladeshi parents of disabled children in the UK: Considerations for service providers and the disability movement. *Health and Social Care in the Community, 11*(6), 502–509.

Chamba, R., Ahmad, W., Hirst, M., Lawton, D., & Beresford, B. (1999). *On the edge: Minority ethnic families caring for a severely disabled child*. Joseph Rowntree Foundation, The Policy Press.

Children and Families Act. (2014). *The stationery office limited under the authority and superintendence of Carol Tullo, Controller of her majesty's stationery office and queen's printer of acts of parliament*. Retrieved December 15, 2020, from http://www.legislation.gov.uk/ukpga/2014/6/pdfs/ukpga_20140006_en.pdf

Croot, E. J., Grant, G., Cooper, C. L., & Mathers, N. (2008). Perceptions of the causes of childhood disability among Pakistani families living in the UK. *Health and Social Care in the Community, 16*(6), 606–613.

Darke, P. (2004). The changing face of representations of disability in the media. In J. Swain, S. French, C. Barnes, & C. Thomapp (Eds.), *Disabling barriers-enabling environments* (pp. 100–105). Sage.

Darr, A., Small, N., Ahmad, W. I., Atkin, K., Corry, P., Benson, J., & Modell, B. (2013). Examining the family-centred approach to genetic testing and counselling among UK Pakistanis: A community perspective. *Journal of community genetics, 4*(1), 49–57.

Department for Education. (2018, January). Special educational needs in England. Retrieved July 26, 2018, from https://www.gov.uk/government/statistics/special-educational-needs-in-england-january-2018

Department of Education & Department of Health. (2015). Special educational needs and disability code of practice: 0 to 25 years. Retrieved December 15, 2020, from https://www.gov.uk/government/uploads/system/uploads/attachment_data/file/398815/SEND_Code_of_Practice_January_2015.pdf

Du Bois, W. E. B. (1897). Strivings of the Negro people. *The Atlantic Monthly*. Retrieved December 15, 2020, from https://www.theatlantic.com/magazine/archive/1897/08/strivings-of-the-negro-people/305446/

Garland-Thomson, R. (2002). Integrating disability, transforming feminist theory. *NWSA Journal, 14*(3), 1–32.

Gorad, S. (1999). 'Well, that about wraps it up for school choice research': A state of the art review. *School Leadership and Management, 19*(1), 25–47.

Griffith, A. I., & Smith, D. E. (2005). *Mothering for schooling.* Routledge Falmer.

Hamdan, A. (2009). *Nurturing eeman in children.* International Islamic Publishing House.

Laura, C. T. (2014). *Being bad: My baby brother and the school-to-prison pipeline.* Teachers College Press.

Moll, L., & Greenberg, J. (1990). Creating zones of possibilities: Combining social contexts for instruction. In L. Moll (Ed.), *Vygotsky and education: Instructional implications and applications of socio-historical psychology* (pp. 319–348). Cambridge University Press.

NAHT. (2018). *Empty promises: The crisis in supporting children with SEND.* NAHT Headquarters.

National Education Union. (2019). Schools at breaking point as funding pressures hit support for SEND pupils. Retrieved December 15, 2020, from https://neu.org.uk/press-releases/schools-breaking-point-funding-pressures-hit-support-send-pupils

Office of the Children's Commissioner. (2014). A rights-based approach to education. Retrieved December 15, 2020, from https://www.childrenscommissioner.gov.uk/wp-content/uploads/2017/07/rights-based-approach-good-education-system.pdf

Pantazis, C., & Pemberton, S. (2009). From the 'old' to the 'new' suspect community examining the impacts of recent UK counter-terrorist legislation. *British Journal of Criminology, 49*(5), 646–666.

Poston, D. J., & Turnbull, A. P. (2004). Role of spirituality and religion in family quality of life for families of children with disabilities. *Education and Training in Developmental Disabilities, 39*(2), 95–108.

Rajchman, J. (1991). *Truth and Eros: Foucault, lacan, and the question of ethics.* Routledge.

Reay, D. (2004). Education and cultural capital: The implications of changing trends in education policies. *Cultural trends, 13*(2), 73–86.

Refugee Action. (2017). *Slipping through the cracks: How Britain's asylum support system fails the most vulnerable.* Retrieved December 15, 2020, from www.refugee-action.org.uk/resource/asylum-support-delays-report.

Rizvi, S. (2015). Exploring British Pakistani mothers' perception of their child with disability: Insights from a UK context. *Journal of Research in Special Educational Needs.* https://doi.org/10.1111/1471-3802.12111

Runswick-Cole, K. (2007). 'The Tribunal was the most stressful thing: more stressful than my son's diagnosis or behaviour': The experiences of families who go to the Special Educational Needs and Disability Tribunal (SENDisT). *Disability and Society, 22*(3), 315–328.

Runswick-Cole, K. (2011). Time to end the bias towards inclusive education? *British Journal of Special Education, 38*(3), 112–119.

Ryan, F. (2018, August 6). 'It's horrifically painful': The disabled women forced into unnecessary surgery. *The Guardian*. Retrieved December 15, 2020, from https://www.theguardian.com/society/2018/aug/06/disabled-women-surgery-catheter-accessible-toilets

Ryan, F. (2019). *Crippled: The austerity crisis and the demonization of disabled people*. Verso.

Shah, R. (1995). *The silent minority-children with disabilities in Asian families*. National Children's Bureau.

Weale, S., & McIntyre, N. (2018, October 23). Thousands of children with special needs excluded from schools. *The Guardian*. Retrieved December 15, 2020, from https://www.theguardian.com/education/2018/oct/23/send-special-educational-needs-children-excluded-from-schools

PART II

CHAPTER 3

Mothering in a Muslim Context

Writing primarily to a Western audience, the title of this chapter has not been chosen lightly. There are negative and dangerous stereotypes about the extent to which the rights and agency of Muslim women shape their identity, family, professional and other important aspects of their lives (Abu-Lughod, 2013). Had the focus been on Muslim mothering in a Western context, the Western reader might assume a homogeneity within the category of 'Muslim women'. However, an archetype of a Muslim family living in Britain does not exist, and nor is "*Islam a place from which one can come*" (Abu-Lughod, 2013, p. 69). I am also aware of the tensions and contradictions perhaps when I refer to the 'Western context' as a fixed crystallised White space, where only those who are White can feel at home. The challenge within home-school literature on minority parenting is how to avoid sensationalising Muslim family experiences of supporting their children without reducing everything to their 'Muslimness', and yet still emphasising that their experiences of rearing children with disabilities occurs at the intersection of their multiple identities and multiple axes of oppression. This dilemma is sadly not limited to Muslim experiences within education; as Abu-Lughod (2013) notes, there is also a default position to how Muslim women are represented within media and academic discourse.

The representation of Muslim communities living in any part of the world seems to emanate from a certain epistemic way of knowing and understanding their experiences; that epistemic way of knowing has only

© The Author(s), under exclusive license to Springer Nature Switzerland AG 2021
S. Rizvi, *Undoing Whiteness in Disability Studies,*
https://doi.org/10.1007/978-3-030-79573-3_3

benefitted those imposing this gaze, and done more harm to those subjected to it. Pappano and Olwan (2016) tell us that the stereotypes of Muslim mothers constructed by the Western gaze often *"portrays them as shadowy, veiled figures in the background, repressed by a violent, domineering patriarchal religious culture and usually, shown as little more than silent appendages of their husbands"* (p. 3). As a result, religion is the definitive factor that seems to explain away all their mothering practices. The media has certainly been at the forefront of scrutinising the extent of the successful integration or otherwise of Muslim mothers into British mainstream society, as I discuss later on in this book. Have they done a good job of raising Muslim children to be British? Are the Muslim mums doing enough to engage with schools?[1]

It would be wholly incorrect, and it certainly is not my intention to state that religion is not significant to the UK's Muslim community. In fact, religion is a strong social signifier for all sections of Britain's South Asian community, not only British Muslims but also British Hindus, British Sikhs and ethnic South Asian Christians (Abbas, 2005). Historically, South Asian migration to Britain has been diverse, having included Hindus, Muslims, Sikhs, Christians and Parsis (Peach, 2006). Whilst 92 per cent of British Pakistanis and British Bangladeshis are Muslim, in contrast British Indians have displayed a marked religious diversity in their composition with 45 per cent identifying as Hindu, 29 per cent as Sikh, 13 per cent as Muslim, 5 per cent as Christian, and a very small minority of Parsis (Peach, 2006). Religion as practiced within the UK's South Asian communities is heterogeneous and fluid in nature; Hinduism contains approximately 3000 different castes, Islam comprises 72 sects and sub-sects, whilst Sikhism has 9 castes. The Census (National Office of Statistics, UK, 2011) reported that Christianity (at 59 percent of the population) was the largest religious group within England and Wales, whereas Islam (at five percent) was the largest minority religion; 25 percent reported no religion affiliation. Whilst much of the home-school literature compiled in the UK during the 1980s and 1990s examined South Asian communities as a heterogeneous whole, however, since

---

[1] The Independent (2014) probed whether Muslim mothers should be trained in computing to help spot radicalisation in their families. https://www.independent.co.uk/news/uk/home-news/muslim-mothers-should-be-trained-in-computing-to-help-to-spot--radicalisation-9040289.html

the events of 9/11 and subsequent terrorist attacks in the West, the focus has shifted from South Asians to British Muslims (Din, 2017).

THE RACIALIZATION OF MUSLIMS

'Muslim' is not in itself a racial category such as 'Black'; however, Muslims have become racialised within Western discourse. Modood (2010) highlights that Jews were originally a faith group, however, an oppressive history of racialisation culminated in them being re-labelled a race. Therefore, when considering critical race discourse we should disregard who falls inside or outside the race category, and focus instead on how certain communities have been racialised over time and how they have become 'suspect' solely because they belong to that community. However, individuals can experience multiple layers of racism as a result of different systems of oppression at play that work to marginalise their multiple positionalities. For instance, calling a Pakistani Muslim a 'Paki' would be classified as ethnicity-based racism which reflects a resentment of Pakistanis, whereas labelling them a 'terrorist' displays hostility and racism towards their 'Muslim' identity (Modood, 2010).

Modood (2010) suggests that hostility towards Muslims living in the West can be traced back to the Satanic Verses affair in 1989. This relates to the publication of The Satanic Verses by British author, Salman Rushdie which was declared a work of blasphemy by Ayatollah Ruhollah Khomeini, and who subsequently issued a *fatwa* or religious edict which ordered Muslims around the world to execute Rushdie. This controversy set the right to free speech, a fundamental entitlement in the West, in direct conflict with a theological Islamic interpretation of blasphemy that prohibits offending Islam, the Prophet Muhammad or indeed other biblical prophets which is sincerely held by millions of Muslims in the West. Prior to this, Islam or Muslims had not entered into public political discourse within the UK, where the debate on immigration was still dealing with 'colour racism'. However, following the terrorist attacks of 9/11, 7/7, and the Woolwich murder,[2] as well as the rise of the Islamic State, have led to exaggerated, demonising and often misleading perceptions. These include Sharia law being applied in the West, the full-face veil, halal slaughter

[2] In May 2013, two men murdered soldier Lee Rigby in Woolwich, London. The murderers were Muslim by faith and admitted to being inspired by Al-Qaeda, an extremist Islamist terror group.

practices, and most recently, the crisis of mainly Muslim refugees in Europe, have all culminated in the perception being entrenched that Islam is incompatible with 'British values'. As a result, this has pressured many moderate Muslims into having to explicitly prove their allegiance to liberal Western values. Pantazis and Pemberton (2009) suggest that Muslims are perceived as the new enemy within, threatening British pluralistic society. They apply Hillyard's notion of a 'suspect community' to illustrate how all British Muslims are now suspect and are subject to surveillance, rather than the minute minority of violent extremists who pose a threat. Pantazis and Pemberton have defined the suspect community as,

> ... A subgroup of the population that is singled out for state attention as being 'problematic'. Specifically, in terms of policing, individuals may be targeted, not necessarily as a result of suspected wrong doing, but simply because of their presumed membership to that sub group. Race, ethnicity, religion, class, gender, language, accent, dress, political ideology or any combination of these factors may serve to delineate the sub-group. (Pantazis & Pemberton, 2009, p. 649)

This has led to a fear of Islam and Muslims generally amongst the wider British public, which has fuelled Islamophobia and anti-Muslim hatred and paradoxically helped Islamist extremists to recruit followers from the British Muslim community. Most mainstream politicians have rejected the link between religion and terrorism; for instance, US Secretary of State, Hillary Clinton stated that the Islamic State was "*neither Islamic or a State*" (Merica, 2014). However, we have also seen politicians like John Denham MP (UK Home Office Minister) state in 2005,

> Few terrorist movements have lasted for long without a supportive community ... certainly most people in it would not want to become personally involved ... whether or not they condone violence they see terrorists sharing their world view, part of the struggle to which they belong. (John Denham cited in McGhee, 2008, p. 69)

Moreover, in 2013 the UK Attorney General, Dominic Grieve MP accused the British Pakistani community of endemic corruption in a newspaper interview, in effect labelling the entire community as 'suspect' rather than only those few corrupt individuals. Some political academics have suggested that there is no evidence proving that Muslims have been officially targeted and victimised (Greer, 2010); however, numerous studies have found that new anti-terrorism legislation disproportionately affects South Asian Muslims. For instance, government figures on 'stop and search' checks by the police in England and Wales from 2006–2007, show that Black and Asian Britons were three times more likely to be stopped and searched under counter-terrorism measures than White Britons (Ministry of Justice, 2008). Mythen et al. (2009), who researched 32 British Pakistani Muslims aged between 18–26 years, explored media representation of Muslims in the UK, and their experiences of victimisation as a result of counter-terrorism legislation. Participants in this research expressed anger and frustration at their experiences of stigma and being bracketed with terrorists, solely because they nominally shared the same faith. This stigma affected their identity and the way they behaved and positioned themselves within the public sphere, viewing themselves through the eyes of others. Many participants expressed how they had to simultaneously prove that they were 'good Muslims' and loyal Britons (Mythen et al., 2009). This study highlights the specific challenges that younger South Asian Muslims face in negotiating their fluid identities against family, community and political pressures. Whilst acknowledging that terrorist attacks are reprehensible with tragic consequences, the participants also believed that the government had used terrorism as an excuse to scrutinise and stigmatise all Muslims rather than engaging with Muslims to address the causes of terrorism. These young South Asian Muslims faced the prospect of integrating into a society which considered them suspect, unwanted and dangerous.

Negative media and public attitudes towards Muslims are more prevalent than some academics acknowledge. The media regularly depicts Muslims as the new suspect community, echoing Pantazis and Pemberton; for instance, The Telegraph newspaper quoted a YouGov survey which had found that 53 per cent of respondents felt "*threatened by Islam*" in general, rather than specifically Islamist extremists.[3]

The UK government's counter-terrorism measures have created mistrust with British Muslims, who rightfully feel increasing levels of suspicion and stigma from the government, the media and wider society after each terrorist attack in the West. Unsurprisingly, such negative attention permeates political leadership on a macro level, as well as the health,

---

[3] YouGov survey was conducted in the UK for The Telegraph newspaper in August 2006 http://www.telegraph.co.uk/news/1527192/Islam-poses-a-threat-to-the-West-say-53pc-in-poll.html

education and employment experiences of British Muslims on a micro level, affecting how they exercise citizenship, whether they acknowledge or resist aspects of their identity, their reactions to socio-political pressures, and their participation in public and social spheres.

ISLAM AND MOTHERING

Given that Islam is a social signifier for Muslims, it is important that we give due consideration to the role that faith plays for British Muslims, and how it shapes their understanding of motherhood and mothering practices in their day-to-day experiences. Whilst there are many sects and sub-sects within Islam, all Muslims follow the basic tenets of Islam described in the Quran (Abdel Haleem, 2004). These tenets are: *Tawhid*, the belief in the Oneness of God and that the Prophet Muhammad is God's final messenger; *Salat*, the establishment of prayers at dawn, noon, mid-afternoon, sunset and at night; *Zakat*, giving charity based on a percentage of your earnings and possessions to the poor and needy; *Fasting* from dawn to sunset during the month of Ramadan and abstaining from sins, and; *Hajj*, a pilgrimage to the Holy city of Mecca in Saudi Arabia once in one's lifetime for those who can financially afford and physically perform it. The Quran also outlines other beliefs that are common to all Muslims, regardless of their sect, including belief in eternal life after death, belief in the revelations of the Old Testament (the Torah) and New Testament, the Psalms and the Quran, and belief in accountability of one's own actions on the Day of Judgement.

Din (2017), who writes on Muslim mothering and the schooling experiences of their children in a UK context, notes that whilst motherhood is discussed in several chapters of the Quran, it is primarily considered around three themes: historical stories relating to specific mothers; legal jurisdiction for inheritance, marriage and divorce; and the status of mothers in the family and social sphere. She notes that the Quran does not prescribe or advocate a particular form of mother; the Quran mentions single mothers, foster mothers, women who became mothers in old age, mothers with infertility problems, and mothers of 'believers'. Whilst her book does not specifically cover the experiences of Muslim mothers of children with disabilities, Din (2017) does provide a nuanced understanding of how inclusive the Quran is in terms of defining mothering and motherhood. The Quran does not favour any one form of mothering over another, nor assign any significance or judgement on the pre-natal behaviour of the mother in relation to her offspring. The religious theological text on mothering that Din (2017) expanded upon seems to completely diverge with what existing studies have reported about the experiences of British Muslim mothers, and how they view their mothering of their children with disability through a religious lens. Whilst the aim of these studies did not specifically outline an examination of the religious understanding of mothering a child with disability, the findings nonetheless present a very static and limited understanding of how the religious lens shapes British Muslim disabled families.

Two studies that explore South Asian familial and religious perceptions of disability to some extent are Bywaters et al. (2003) and Croot et al. (2008). Bywaters et al. (2003) conducted a qualitative study with 19 families, as part of an advocacy project with British Pakistani and British Bangladeshi families of children with severe learning disabilities (SLD) in Birmingham. Parental explanations for their children's disability varied between biomedical and religio-cultural views. Interestingly, the religious views of families that were reported were not negative, but rather were limited to how disability was simply a fact that was accepted as God's will. In the same study, however, two mothers did report instances of religious policing within their respective communities, which blamed their children's disability on the mothers' supposed sins; nonetheless, both those mothers repudiated this type of 'mother blaming' using specific medical explanations for their child's disability. It appears that Islam is again presented as a passive factor in the study by Bywaters et al. (2003), in revealing that the mothers accepted their child's disability as either God's will, or merely as a nuisance when they experienced religious or community policing, both of which could be dealt with and tackled using biomedical explanations. Whilst Bywaters et al. (2003) did negate existing stereotypes about Muslim families who define disability in both medical and religious terms, by reporting that religious explanations of disability did not preclude mothers from striving for the best possible support from service providers, nonetheless, the study approached this issue from the premise of evaluating whether Islamic teaching is compatible on a functional level with an understanding of scientific practices. Interestingly, the study did highlight that British Muslim parents did not attribute disability to social attitudes or external structural barriers within society, suggesting that the social model of disability is too Eurocentric to work in an ethnic minority context.

In the study by Croot et al. (2008), there was relatively greater exploration of the religions lens. Croot et al. (2008) engaged two Pakistani parents and a support group worker to design and implement their qualitative collaborative research. Ultimately, the study sample entailed interviews with nine families of disabled children, including seven families from Pakistani backgrounds. All participants held religious explanations for their child's disability but they varied in how they applied their religious understanding, although some couples did hold divergent religious frames. Notably, all parents regarded disability in a generally positive light, whether anticipating that a 'guaranteed passage to heaven' awaited them as a reward for parenting their disabled child or simply the everyday pleasure their child brought them. Some parents considered that raising a disabled child was a divine test of their faith and compassion, or that they had been 'chosen' by God for this role. One mother who had previously questioned whether her child's disability may have directly resulted from her past sins, however, later realised that disability was beyond her control (Croot et al., 2008). Two other mothers also revealed that their extended families and community members had also attributed their child's disability to a curse by evil spirits; significantly, the mothers repudiated this assertion, suggesting that ignorance led to a poor awareness of disability. In the study by Croot et al., we see that the authors have attempted to show how Muslim mothering of children with disability does not occur in binaries, and that mothers can simultaneously express biomedical and religious explanations. However, both studies were predicated on the assumption that the religious views reported by the sampled parents were unique to parents of disabled children, and lacked a comparison with responses from Muslim parents of able-bodied children.

Whilst I was speaking with the mothers in my study, the subject of religion was never far from the conversation. It was present whether we were discussing politics, the mothers' exhaustion from their household duties and other unpaid labour, their mental health, or their children. However, my central assertion is that the mothers did not only hold religious views of these issues alone, rather, that their experiences as British South Asian Muslims imbued their entire lives with a religious lens. In the sections below, I have grouped my conversations with my research participants around three broad themes.

- Religious Citizenship (rights, obligations)
- Religious Interpretations
- Religious Inclusion (in Muslim circles, in Secular society)

RELIGIOUS CITIZENSHIP

In our conversations, I wanted to probe the extent to which able-bodied Muslim mothers went about instilling the values of Islam and of being a 'good Muslim' in their children with disability. Do they have the same concerns for mothering their other non-disabled children? Our conversations were a medley of emotions for these mothers, who wanted to assert their religious knowledge and state that there is freedom and reason within Islam which absolves disabled people from ritual obligations such as reciting daily prayers or fasting during Ramadan, whilst at times infantilising the agency of their child with disability with regard to developing their own sense of being a Muslim. There was an acute awareness of the physical labour involved in being a practicing Muslim, such as the physical act of performing ritual purification before reciting obligatory prayers five times a day, as well as of knowing that mothering children with disabilities requires a new ontological imagining of what it means to be a Muslim in a non-able-bodied way.

> It's in Surah [Quranic chapter] Muhammad I think, it says that under Shariah [Islamic law], religious obligations are not mandatory for those children who I'd describe as special … they don't comprehend or understand about Islam or religious obligations, so how can you say you want to teach them about religious obligations? So for example, namaz [daily prayers] won't be mandatory for them and fasting won't be mandatory when they don't have an understanding, so how can they learn to do it? … first of all, there's no compulsion in Islam just like all other religions so with special needs people who don't have the understanding then Shariah isn't binding on them … There's no need to consult or even discuss about whether Daniel will learn or not when he doesn't even have a basic understanding of life, I mean how can you teach him?—Maham

> His religious responsibilities? He doesn't have any because religious responsibilities are only binding on those who have got some understanding … you can't put religious obligations on these kinds of children. Why am I saying I can relate to it? Because I've said the same things about Aamir. He's got no understanding of anything, not even danger. He could put something in his mouth and he can choke on it … he's got no understanding of how something that's hard can hurt him, so how can he understand religion? A 14-year-old is obliged to pray five times a day but, you know, he won't be able to recite it and understand it. So with Aamir, he's under no obligation.—Maria

There were no commonalities between Maria and Maham's sons in terms of their disability; Maria's son, Aamir is a 14-year-old who has a degenerative metabolic condition whilst Maham's son, Daniel is a 7-year-old who has global developmental delay with autistic features. Nonetheless, both Maria and Maham seem to have implicitly understood that neither Aamir nor Daniel have an obligation to perform the ritualistic aspects of Islam, such as performing daily obligatory prayers, fasting during Ramadan, or performing the Hajj pilgrimage to Mecca. It is incumbent on Muslims to be in a state of constant purification and cleanliness (ghusl) when performing these fundamental acts,[4] as well as having a belief in and an understanding of the basic tenets of Islam such as a belief in God, the prophethood of Muhammad, or belief in life after death. This was a situation where I felt the tension between maternal agency and the child's rights to practice their own religious citizenship. For both Aamir and Daniel, being a Muslim meant being raised within the practices of a Muslim household such as eating halal food, celebrating the Eid festival and other religious events, as well as being raised in a household that values the Islamic notions of faith, modesty and familial obligations. However, claiming that Islam is reasonable and inclusive for all Muslims is a contradiction because, on the one hand, although it does provide some individuals with the flexibility to practice the faith in a way that is compatible with their multiple identities and positionings, yet on the other hand, the rights to full citizenship within the faith are at risk of being denied to those who are not able-bodied. So whilst there is an inherent humanitarian benevolence within Islam with regard to disabled people as the mothers highlighted, yet at the same time they are *expected* to feel grateful that they have been included.

The circumstances would be different if disabled individuals were fully agentic in determining which elements of religious practice are accessible and compatible for them and which parts of the faith they want to let go, as opposed to this decision being made for them as an act of leniency and compassion, which in this case was made by the mothers. Religious identity for these young children is very different to the religious identity held by their able-bodied mothers. The maternal understanding of their children's religious identities is almost apolitical, and assumes a docility and

---

[4] These acts become invalid with passing of urine, faeces, gases, and bleeding. Individuals with incontinence issues have more relaxed rules around the extent to which they can maintain a state of ghusl.

submissiveness that removes the children's agency from this whole process. As Maham states in the interview, "*There's no need to consult or even discuss about whether Daniel will learn or not when he doesn't even have a basic understanding of life, I mean how can you teach him?*" This is complicated further because the children's ethnic and cultural identities may have closer similarities with their able-bodied mothers, even if the boundaries between cultural and religious practices have been blurred; for instance, an insistence on maintaining a traditional dress code such as wearing kurta shalwar, because of religious beliefs around modesty alongside simultaneously held cultural beliefs around morality and sexual rights. The research by Islam (2008) with 13 British Pakistani and British Bangladeshi young people with disabilities seems to reflect this disjunction, in that some of the disabled young people in this study did not identify with their parent's religious heritage even though they shared similar ethnic and cultural roots. Islam's (2008) study does show the limitations of applying a Eurocentric social model of disability to understanding the religious citizenship of communities from the Global South that relies on an individual's 'able-bodiedness'. While her research does briefly touch upon how an impairment or disability could mediate adherence to religious practices, it does not unpack the early socialisation of children or young people with disability into understanding what it means to be a Muslim, how much of it relies on being able-bodied, and whether they could be a practicing Muslim given that they are disabled (Islam, 2008).

Another possible explanation why a child's agency may be absent from their religious identity is that, in some instances, mothers were actively performing the religious rights on behalf of their child. Islam contains a provision which permits the participation and performance of certain obligatory acts by a proxy; for instance, one can give zakat (obligatory charity) or fast on behalf of someone else. In the examples I quoted above, both Maham and Maria were performing religious acts on behalf of their children, primarily because they wanted their child with disability to have an active Muslim identity. It is also worth noting that those acts which did not require additional physical labour such as giving charity, were more readily performed compared to acts that inherently did entail physical effort such as reciting daily prayers or fasting. I did not ask for clarification on this issue because asking this could have been perceived by the mothers as a judgement of their maternal practices and decisions. The mothers' awareness that their children did not need to fulfil any religious obligations meant that they felt less pressured to publicly display their child's religious conformity within British Pakistani Muslim spaces, and were more flexible in how they parented their child in a religious context. The mothers also felt more enabled in transmitting their own religious values due to their greater religious knowledge.

> ... your first priority is your child so culturally your child needs to learn Islamic teachings and follow Islam ... but if he's not able to understand it, you can't force him and say he's got to know this by this certain age because, even with a normal child, each one develops at his own pace, some will develop and understand things early whereas others might take more time. So I think it's down to you, as a parent, to know if your child's able to manage and understand that.—Shehnaz

Often in Muslim communities across the world, children are expected to attend a madrassah (religious school) every weekend and to become familiar by a certain age, often before puberty, with various aspects of the religion; for instance, they learn about various prophets, short stories from the Quran, cultural values that might have some mention in the Quran or the hadiths (historical narrations about the Prophet Muhammad), as well as learning the Arabic alphabet that will help the children to read the Quran when they are older. For Shehnaz, all such aspects that are considered typical of a Muslim upbringing were, in her children's case, subject to her own discretion. Shehnaz's children, Amna and Tariq, both of whom are disabled attended religious classes at weekends to help them to learn to read verses from the Quran, however, the pace at which they were learning and engaging with Quranic texts was visibly slower than Shehnaz's able-bodied children. She told me that she had personally explained to the imam (religious teacher) who taught the classes that both Amna and Tariq required a more inclusive way of teaching the Quran. To her, religious participation by her children with disability had to be at their own pace and built around their specific needs, rather than according to the cultural expectations of age-related religious milestones. The mothers in my study were aware that religious practices were often enacted within the cultural framework of the South Asian community, and it was that community pressure which often compelled their conformity to religious practices rather than direct theological religious doctrine. For instance, Saira and Shehnaz both indicated that constant public engagement within the community such as regularly attending religious school and the mosque might be too much for disabled children, and they would not put their child under such pressure.

Whilst having a disability meant that their children were exempt from certain religious obligations, it also relieved the mothers from worrying about their children being exposed to those aspects of British culture which conflicted with their Islamic values. For instance, this helped to reduce the exposure to mainstream cultural choices that young able-bodied people in the West would typically face, such as having a girlfriend or boyfriend, or socialising in nightclubs or pubs. This was a contentious issue that is further examined in later chapters; nonetheless, it was quite evident that the mothers' religious and cultural values were less in conflict with their child's gender and sexual rights, and that this reduced level of conflict was in part mediated by their child's disability. For some mothers like Kiran and Maria, their child's disability actually appears to have helped to preserve their Islamic faith and cultural values across generations because their beliefs were not questioned by their child with disability in the same way that an able-bodied child would have challenged. Moreover, some mothers compared their experiences of mothering their child with disability with their other abled-bodied children, and rationalised and highlighted what they perceived as the positive experiences of caring for and supporting a child with disability to downplay and mitigate the difficulties they experienced on a daily basis. However, it seems that some mothers did infantilise their children, rationalising their overly protective behaviour by citing documented cases of sexual abuse of disabled children and young people. They utilised their religious identities as a shield to protect their children from potential abuse, not acknowledging that incidents of sexual abuse were only more visible in this country because it was more openly discussed and more widely reported as a result.

RELIGIOUS INTERPRETATION

There has been some scholarship into the religious interpretation of disability in the Quran. Although there is no direct reference to the term 'disability' within the Quran, however, there are references about the blind, deaf, weak and lame. And in similarity with other Abrahamic texts, there is some contention between the importance of the literal meaning versus an interpretation of the text. For instance, the Quran contains the following passage,

Yet, verily, it is not their eyes that have become blind—but blind have become the hearts that are in their breasts! (Quran, Surah 22, Verse 46)

In this context, religious academics (Bazna & Hatab, 2005; Ali, 1996) have suggested that the quoted text refers to people who have closed their minds to spiritual knowledge as opposed to people who have a physical loss of vision. Other religious and disability activist academics (Turmusani, 2001; Barnes, 1992) would outright disagree with and reject the ableist language used within the Quran, the Bible and other religious texts. They have also repudiated the analogy of disabled people being individuals who have been punished or condemned for their supposed sins, and who have mental health needs or have become physically disabled as a consequence (Turmusani, 2001; Barnes, 1992). Interestingly, references to illness within the Quran have been done from a rights-based and caring perspective, or they portray illness as symbolic of faith and perseverance. Miles (2002) argues that interpreting these religious texts in a literal sense is not helpful because it disregards their metaphorical frames of reference. Rather, these texts should be interpreted in accordance with present day context, and in consideration of the intersecting factors that mediate the experiences of these faith communities. This argument seems to make sense as one engages with scholarship on the religious understanding of disability and on the role of religion in parenting children with disability (see the earlier discussion on the studies by Croot et al. 2008 and Bywaters et al. 2003). What Miles (2002) does not consider is how knowledge of these religious texts and understanding of disability which has been built over decades is preserved and passed on within communities, and how it intermeshes with patriarchal structures within South Asian Muslim communities across the UK to often incite mother blaming and influence mothering roles.

In choosing to raise their children with disability as Muslims in their own right, meant that over time the mothers in my research gained a very good understanding of religious jurisprudence. This knowledge of religious jurisprudence meant that the mothers were able to exercise their own agency in performing their caring and mothering roles. It also meant that they did not have to rely solely on one form of expertise (i.e. medical), but rather were able to utilise a range of expertise from medical, social, educational and religious disciplines to make informed decisions. Although a few studies (Skinner et al., 1999; Poston & Turnbull, 2004; Graybill & Esquivel, 2012) have explored the role of religion, and have highlighted the positive effects of religiosity and spirituality on the mental health of parents of disabled children, however, the role of religion within parental involvement and home-school literature is absent. This exclusion of religious belief not only dismisses the positive role that religion can play within home-school relationships, but it also marginalises certain aspects of identity which are important to particular groups given their exclusion from mainstream support services. Religion also provided a positive and uplifting framework for mothers to lean on, providing coherence to their experiences as first-time mothers of disabled children. Indeed, I posit that the mothers' religious lens can be viewed as an alternative framework to the traditional Western Kubler-Ross' (1969) Model of the Stages of Grief,[5] because the mothers in my study made sense of their experiences of mothering a child with disability through their religious faith. The Kubler-Ross Model (1969) provides a very narrow ableist psychological lens to understanding the parental response to learning that their child has been diagnosed with a disability; it may not reflect the sentiments of those parents who do not view a disability diagnosis as a 'grieving' event. The mothers delved into Quranic teachings in order to gain knowledge about what their obligations were in relation to mothering and caring for their child with disability, as well as to gain a broader perspective on their mothering experiences.

> I think the more and more you go into your religion, the more patience you get. I feel that no matter how stressed you are, when you sit down, the Quran has got answers to everything, that's our deen (religion), that's what we believe in ... you can connect it to what's happening in your life at that time.—Maria

For other mothers, religion provided a meaningful narrative of their past, present and future, which allowed them to reformulate their notions of mothering. This is exemplified in Maham's comment,

> It's God's will and we thank God for him [Daniel], but when we think how difficult the world is for him then our responsibilities are much greater ... God has made me a mother and huge responsibility comes with that.—Maham

---

[5] Elisabeth Kubler-Ross (1969) identified and codified five emotional stages experienced by survivors of a close intimate's death, that is, denial, anger, bargaining, depression and acceptance.

For Maham, like many mothers in my research, the notion that her mothering role had been ordained by God helped her to recuperate, represent and rationalise her experiences of supporting her child with disability. This type of attitude to religion has also be reported in other communities where religious faith is strong. For instance, the USA-based study by Skinner et al. (1999)[6] found that maternal responses about their children's disability and their resulting responsibilities were deeply embedded within their Catholic faith and its interpretation of what being a 'good mother' entails. Catholicism extols the Virgin Mary as the epitome of the ideal mother, who is both devoted and self-sacrificing. The mothers rationalised their child's disability in the belief that God had chosen them as being capable of mothering a disabled child. It can also be argued that the fact that Maham was living within the patriarchal confines of a fairly traditional South Asian joint family system, meant that she may have more willingly accepted and more naturally embraced her role as the forbearing and dutiful mother-caregiver (as well as deferential wife and daughter-in-law). Nonetheless, some mothers did also report expressing anger during the early disclosure stage of their child's disability.

> It was so emotional for us when he [Ahmed] was born. We asked, why us? Why him? You go through that, but again, that's what was meant to be … that's God's way … so God tests people in different ways, and this is a test … so religiously we go down that route and it makes it bit easier for us.—Kiran

The mothers in my study did not identify with the Kubler-Ross Model, and instead adopted an alternative trajectory to explain their experiences which was set within a religious narrative of successfully overcoming divine tests and of being content with God's will. Interestingly, for some mothers, a state of submitting to and accepting God's will could simultaneously coexist with being angry with God in the same moment. Expending labour into parenting a child in order to earn a place in a paradise has long been a cherished belief held by all practicing Muslim parents, regardless of whether or not the child has a disability. However, in addition to this, some mothers reported that a relationship which requires a continual performance of labour, such as caring for a child with disability, effectively

---

[6] The study by Skinner et al. (1999) explored the narratives of 150 mothers from Latino American families with regards to their experiences of motherhood.

guaranteed them a more elevated place in the hereafter; such a belief enabled mothers such as Maham to embrace their relationship with God on more just terms.

Mothers like Kiran found another more personal way of channelling the daily toil and frustrations associated with mothering a child with disability, by verbally expressing anger at God at various stages of her child's life rather than losing the connection with her religious faith altogether. This form of anger reflects an authentically raw relationship with God, one that demands of God to fulfil His promise as a just and merciful Supreme Being; after all, one cannot make demands of or express anger at a higher being, if one is not first and foremost in a spiritual relationship with God. Kiran and Alina's anger is similar to the response of Gottlieb (2002), a Jewish father of a 14-year-old girl with multiple disabilities.

> And as for God? I know that for a start, anger at God is a legitimate aspect of spiritual life. To be angry at God is just as much a part of prayer as is love or devotion, awe or repentance. In all those states of mind, feeling and soul, we are confirming that moral laws bind the universe as a whole, just as they bind us as moral agents within it. That's why I think I'm entitled to rage at God when I witness Esther's pain, her limitations and disappointments. When I think that "normal" love relationships and opportunities will simply never be hers. As she herself said, when I told her that because of her special needs she could not return to her beloved day camp as a counsellor: "Now's the time to scream at God." (Gottlieb, 2002, p. 11)

Anger by able-bodied parents of disabled children can sometimes be misunderstood as anger towards the child; therefore, some may argue that perhaps anger should not be expressed in public spaces in a culture where parent blaming is already a part of professional discourse. There is also an ethical duty of being an ally as a parent and displaying anger could be perceived as ascribing to an ableist lens. As raised in the Chap. 1, are we unfairly holding motherhood to a higher standard of disability justice than the oppressive systems and policies that marginalise both the mother and child? How should Kiran's anger be interpreted? More importantly, who ultimately gets to judge her anger? Is she mourning the loss of 'a perfect child' that has been obsessed over within special needs literature? Or is her anger a part of her lived experiences, rather than merely a stage within a model of grief? For Kiran and Alina, who both bear witness to the physical exclusion of their sons on a daily basis, their anger is a part of their justified spiritual response to God. Importantly, their rage is not projected at their sons, but rather at those activities and events that they would like their sons to participate in but are unable to.

As I discussed earlier, the Sunni school of thought within Islamic theology has explored the religious beliefs surrounding parenting responsibilities. For the mothers in my study, having a disabled child attested to the fact that it was God's will to put them through a series of arduous trials, not to punish them but to make them realise that they were capable of looking after a disabled child. The mothers' in-depth understanding of their religious faith also helped them to tackle any negative religious explanations of disability that were offered by community members or by their extended families. Some mothers recalled how they had been told that their child's disability was a punishment from God for their sins; however, they counteracted this by asserting that God was not vengeful and certainly would not punish an innocent child for any supposed sins committed by a mother.

Maternal religious knowledge also challenged those professionals coming from the perspective of the medical model of disability, who viewed the mothers and their children through a deficit lens. Many mothers commented that the medical prognosis given by professionals was not the final verdict in relation to their child's disability, and that these medical experts were not God,

> Yes, I got very upset and angry at the doctor, and I said to him 'Who made you God?' You know life and death is in the hands of God, no human can say that. I said I can go across the road and get knocked down by a bus and die tomorrow but you can't predict that … so how can you say he's [Aamir] only going to live for two more years?—Maria

Maria felt angry and cheated by the doctors after Aamir was diagnosed late, rejecting Aamir's medical prognosis and preferring to place her faith in God. With Shehnaz, doctors had initially stated that her son, Tariq would be unable to walk; however, with considerable support from Shehnaz and intensive physiotherapy Tariq was able to become independent of his wheelchair. This does not mean that the mothers deliberately chose to believe in a subjective valued-based phenomenon such as religion when they rejected the doctor's science-based medical prognosis. Rather, I would argue that the mothers' religious faith stimulated and energised them to persevere and seek a second opinion, as well as being more open to alternative interventions. This shifts the focus from viewing religion as a primitive and irrational superstition which is the opposite of the scientific and psychological lens applied within disability, and instead considers religion as an alternative parallel paradigm through which mothers who are members of faith communities can make sense of their child's disability. Studies such as Hussain et al. (2002), Jegatheesan et al. (2010), and Michie and Skinner (2010) which have touched on parental explanations of their child's disability as an act of divine punishment or karma, do not propose a way out of such negative beliefs. However, mothers in my study countered such negative perceptions by adopting a stronger religious identity based on a deeper theological knowledgebase in order to challenge those indulging in mother blaming.

Religious Inclusion

Recent studies by Howarth et al. (2008), Poston and Turnbull (2004), and Hussain et al. (2002) which examined the levels of participation within faith communities by disabled families, reported that parents felt disheartened when community participation by their child with disability was not supported by community members; notably, this was a common finding across all religions. In my study, some mothers reported first-hand experiences of exclusion by community members when their children with disability accompanied them to the mosque or to other religious gatherings, whilst other mothers highlighted the unsupportive, indifferent and generally exclusionary attitudes of community members towards disabled families. Moreover, mothers of children with behavioural needs were distressed to report that people at the mosque and other public religious settings were unsympathetic and intolerant of their children, which they attributed to their child's disability being less visible compared to other physical types of disability. To counter their children's exclusion, it was positive to note that many mothers narrated one specific hadith or story about the Prophet Muhammad in which he promoted religious inclusion.

> There's one hadith I have in mind. One very hot day, the Holy Prophet was giving a sermon in the mosque when a woman who was not of sound mind called to him from outside the mosque. He left his sermon and went outside to listen to what she had to say. That women was distressed and talked to him for a very long period. The Holy Prophet did not interrupt the conversation or leave until she had left. When he went back into the mosque to

finish his sermon, the people in the mosque asked why he had spent such a long time outside in the scorching heat to listen to woman who wasn't even of sound mind. They reiterated that no one listens to her because she doesn't know what she's saying. The Holy Prophet replied that, *'If all of Mecca doesn't listen to her, and then if I didn't listen to her today, then who would have listened to her?'*—Parveen

Holding an accurate and detailed religious understanding meant that mothers were able to separate religion from their community's cultural practices, often questioning the lack of theological grounds for their children's religious exclusion from the mosque. This is most evident in Maria's comment,

We used to take him [Aamir] … we've also had complaints when there were speeches going on. We used to take Aamir and people used to get disturbed about it and so my husband stopped taking him. I was very cross about this decision … the ladies' section is upstairs, and I can't carry Aamir upstairs. I tried to, once or twice, I took him because I wanted him to get a feel of the environment because we're Muslim and we go to the mosque. But my husband stopped taking him because other people would get disturbed and they said, 'oh we can't hear, we can't understand'. And so we stopped taking him which is very unfair. We should've put up a fight and continue to take him regardless of what people said, because he's our family … He's a Muslim and he's got as much a right as anyone to be there.—Maria

Such maternal experiences not only confirm the findings of previous studies, but also pose several questions about the role that religious organisations can play in the lives of disabled families. What role can religious organisations play in instilling and developing positive attitudes towards disability amongst community members? This is an important point to consider for ethnic minority communities with strong religious identities, who have a responsibility to welcome all members. It also questions the wider supporting role that faith-based organisations can offer disabled families, such as having information leaflets about different types of disability within mosques which can help to educate members and attendees, or providing disabled parking spaces or wheelchair access as Parveen suggested; meeting the social, physical, practical and religious needs of disabled families would inevitably increase and improve their participation. Selway and Ashman (1998) cite one study which found that 21 per cent of the state budget for disability programmes in Queensland, Australia was allocated to 57 religious groups. Admittedly, this is not a recent statistic, however, it does suggest that faith-based organisations can often act as service delivery points for disabled families; therefore, there is a need to assess the awareness, aptitude and levels of training of religious organisations in supporting the religious participation of disabled families.

Due to their own experiences of community exclusion, I found that many mothers in my study displayed less attachment to their local communities, and expressed greater dissatisfaction with the opportunities available to exercise their citizenship rights. Despite the hostility from faith-based organisations towards their disabled children, the mothers did appreciate the positive role that religion played in their lives.

REFERENCES

Abbas, T. (2005). Recent developments to British multicultural theory, policy and practice: The case of British Muslims. *Citizenship Studies, 9*(2), 153–166.

Abu-Lughod, L. (2013). *Do Muslim women need saving?* Harvard University Press.

Ali, A. Y. (1996). *The meaning of the holy Qur'an.* Amana Publications.

Barnes, C. (1992). *Disabling imagery and the media: An exploration of the principles for media representations of disabled people.* BCODP.

Bazna, M. S., & Hatab, T. A. (2005). Disability in the Qur'an: The Islamic alternative to defining, viewing, and relating to disability. *Journal of Religion, Disability & Health, 9*(1), 5–27.

Bywaters, P., Ali, Z., Fazil, Q., Wallace, L. M., & Singh, G. (2003). Attitudes towards disability amongst Pakistani and Bangladeshi parents of disabled children in the UK: Considerations for service providers and the disability movement. *Health and Social Care in the Community, 11*(6), 502–509.

Campbell, F. K. (2020). Indian contributions to thinking about studies in Ableism: Challenges, dangers and possibilities. *Indian Journal of Critical Disability Studies, 1*(1), 1–19.

Croot, E. J., Grant, G., Cooper, C. L., & Mathers, N. (2008). Perceptions of the causes of childhood disability among Pakistani families living in the UK. *Health and Social Care in the Community, 16*(6), 606–613.

Din, S. (2017). *Muslim mothers and their children's schooling.* Trentham Books, UCL IOE Press.

Gottlieb, R. S. (2002). The tasks of embodied love: Moral problems in caring for children with disabilities. *Hypatia, 17*(3), 225–236.

Graybill, A., & Esquivel, G. (2012). Spiritual wellness as a protective factor in predicting depression among mothers of children with autism spectrum disorders. *Journal of Religion, Disability & Health, 16*(1), 74–87.

Greer, S. (2010). Anti-terrorist laws and the United Kingdom's 'suspect Muslim community': A reply to Pantazis and Pemberton. *British Journal of Criminology, 50*(6), 1171–1190.

Howarth, J., Lees, J., Sidebotham, P., Higgins, J., & Imtiaz, A. (2008). *Religion, beliefs and parenting practices.* Joseph Rowntree Foundation.

Hussain, Y., Atkin, K., & Ahmad, W. (2002). *South Asian disabled young people and their families.* Joseph Rowntree Foundation, The Policy Press.

Islam, Z. (2008). Negotiating identities: The lives of Pakistani and Bangladeshi young disabled people. *Disability & Society, 23*(1), 41–52.

Jegatheesan, B., Miller, P. J., & Fowler, S. A. (2010). Autism from a religious perspective: A study of parental beliefs in south Asian Muslim immigrant families. *Focus on Autism and Other Developmental Disabilities, 25*(2), 98–109.

Kubler-Ross, E. (1969). *On death and dying.* The Macmillan Company.

McGhee, D. (2008). *The end of Multiculturalism? Terrorism, integration and human rights.* McGraw-Hill Education.

Merica, D. (2014, October 7). ISIS is neither Islamic nor a state, says Hillary Clinton. *CNN.* Retrieved December 15, 2020, from https://edition.cnn.com/2014/10/06/politics/hillary-clinton-isis/

Michie, M., & Skinner, D. (2010). Narrating disability, narrating religious practice: Reconciliation and fragile X syndrome. *Intellectual and Developmental Disabilities, 48*(2), 99–111.

Miles, M. (2002). Disability in an eastern religious context: Historical perspectives. *Journal of Religion, Disability & Health, 6*(2–3), 53–76.

Ministry of Justice. (2008). *Statistics on race and the criminal justice system—2006/07.* Ministry of Justice. Retrieved December 15, 2020, from www.justice.gov.uk/docs/race-and-cjs-stats-2006.pdf

Modood, T. (2010). *Still not easy being British- struggles for multicultural citizenship.* Trentham Press.

Mythen, G., Walklate, S., & Khan, F. (2009). 'I'm a Muslim, but I'm not a terrorist': Victimization, risky identities and the performance of safety. *British Journal of Criminology, 49*(6), 736–754.

Office for National Statistics, UK. (2011). *Ethnicity and national identity in England and Wales 2011.* Retrieved June 7, 2018, from http://www.ons.gov.uk/ons/dcp171776_290558.pdf

Pantazis, C., & Pemberton, S. (2009). From the 'old' to the 'new' suspect community examining the impacts of recent UK counter-terrorist legislation. *British Journal of Criminology, 49*(5), 646–666.

Pappano, M. A., & Olwan, D. M. (2016). *Muslim mothering: Global histories, theories, and practices.* Demeter Press.

Peach, C. (2006). South Asian migration and settlement in Great Britain 1951–2001. *Contemporary South Asia, 15*(2), 133–146.

Poston, D. J., & Turnbull, A. P. (2004). Role of spirituality and religion in family quality of life for families of children with disabilities. *Education and Training in Developmental Disabilities, 39*(2), 95–108.

Selway, D., & Ashman, A. F. (1998). Disability, religion and health: A literature review in search of the spiritual dimensions of disability. *Disability & Society, 13*(3), 429–439.

Skinner, D., Rodriguez, P., & Bailey, D. B. (1999). Qualitative analysis of Latino parents' religious interpretations of their child's disability. *Journal of Early Intervention, 22*(4), 271–285.

The Qur'an (M. A. S. Abdel Haleem, Trans.). (2004). Oxford University Press.

Turmusani, M. (2001). Disabled women in Islam: Middle eastern perspective. In W. C. Gaventa & D. L. Coulter (Eds.), *Spirituality and intellectual disability* (pp. 73–86). The Haworth Pastoral Press.

CHAPTER 4

Magnifying Gender and Sexual Rights

The notion that the labour required in performing care and the duties traditionally ascribed to mothers are both gendered is an age-old concern, and is a cause for feminist struggles. This issue, however, has mainly been presented from the position of the liberation of White women. It has been posited by bell hooks (2000) that Black mothers and mothers from other minoritised communities did not locate the locus of their oppression to motherhood and caring. Unlike their White counterparts, they were always part of a capitalist workforce working outside their homes, whether this was in factories, fields or in the homes of White families. Black and minoritised mothers located their locus of oppression within white supremacist patriarchal institutions that forced them to be separated with their families and children. Not surprisingly, in the last few decades as more middle-class White women have entered the workforce, they have realised that inherent alienation, sexism and gender inequalities in the workplace have forced them to make sacrifices and choose between a career and personal growth. As a consequence, bell hooks (2000) argues that motherhood is no longer the enemy and that feminists are increasingly redefining motherhood and raising concerns as to why there is little economic value attached to performing mothering and child-rearing. The revival of interest in mothering from 'mainstream' feminist scholars has had positive as well as negative implications. On the one hand, motherhood is no longer located within heterogeneous nuclear households, as more single women and LGBTQ+ women challenge this narrative. On the

S. Rizvi, *Undoing Whiteness in Disability Studies*,
https://doi.org/10.1007/978-3-030-79573-3_4

other hand, the redefinition of motherhood by middle-class White women has also led to romanticising of mothers as natural caregivers who enjoy a special bond with their child, all of which detaches itself from the pressing question of practicalities of care and reinforces the principles of male supremacist ideology (bell hooks, 2000). She argues that this redefinition of motherhood has again disproportionately allocated the labour of child-rearing to female parents, rather than sharing this labour equally between male and female parents. Even when men seem to be performing effective parenting, they are immediately seen as being maternal or performing mothering, as opposed to being seen as good fathers. Bell hooks calls for the complete ridding of gendered roles within parenting and moving towards shared child-rearing practices, using local community and state funded centres as a way to revolutionise parenting within minoritised communities. However, this does present limitations to those communities who view mothering as gendered and 'special' through a religious lens. To assume that religion, in particular Islam, is paternalistic and oppresses women by imposing the sanctity of mothering on them would again risk viewing this conundrum through a *West* versus *Islam* lens. As discussed in the earlier chapter, Islamic theology does not favour any one type of mother over another; however, it does attribute an elevated status to mothers over fathers. For instance, whilst the Quran does not provide any distinction between male and female parents over child-rearing responsibilities, there are several *hadiths* attributed to the Prophet Muhammad which suggest that mothers occupy an elevated status compared to fathers.

> A man came to the Prophet and said, 'O Messenger of God! Who among the people is the most worthy of my good companionship?' The Prophet said: 'Your mother.' The man said, 'Then who?' The Prophet said: 'Then your mother.' The man further asked, 'Then who?' The Prophet said: 'Then your mother.' The man asked again, 'Then who?' The Prophet said: 'Then your father.' (Bukhari, 1966)

> Your Heaven lies under the feet of your mother. (Sunan an-Nasa'i)

Such religious importance specifically attributed to mothers does feed into the 'special bond between mother and child' narrative that bell hooks warns us about. It is important to note that all recognised orthodox Islamic texts have been written by men, which often reinforce the patriarchal construction of motherhood and exclude women's rights to owning the process of knowledge creation regarding her own roles, responsibilities and agency (Cheruvallil-Contractor, 2016). These religious scholars have theorised and theologised on how pressing issues that women face should be dealt with according to Islamic jurisprudence without embodying those experiences. Nevertheless, my interviewing experiences included many mothers who were happy to perform the role of mothering and yet were worried about how sustainable this arrangement would be as their child approached adulthood. *How does one go about revolutionising parenting when Muslim mothers themselves see this role as their gateway to heaven?* And more importantly, *how does it affect their child-rearing practices and the extent to which their spouses are willing to co-parent?* In interviewing mothers in this study, it was very difficult to separate the threads of cultural patriarchy from the religious lens of viewing mothering on a pedestal. As a critical feminist researcher who operates from a Black feminist lens, I saw it as almost antithetical to question the special status that some mothers believed they hold even if that resulted in them bearing the burden of care for their disabled children entirely on their own.

Religion was the only social institution that validated and valued their labour, whereas other institutions such as health, education and welfare seem to devalue the labour involved with mothering their children. In the context of mothering children with disabilities from minority communities, this debate also needs to be understood and approached intersectionally. As Gabel suggests, *"Mothering is situated within the institution of motherhood and numerous other structures or institutions (for example, class, race, marriage, patriarchy)"* (Gabel, 2018, p. 554). Mothering is also grounded in ableism in that it is valued only in relation to socially valued children. Mothers of disabled children know that *"their children are routinely and persistently denied entry into the category of fully human"* (Runswick-Cole & Ryan, 2019, p. 6). Mothering of children with disability is kept separate, or certainly at the periphery of mainstream mothering discourses. Mothers of disabled children are restricted from joining conversations about what a typical day of mothering entails because they fail to conform to coming-of-age rituals such as attending football matches, taking school examinations, proms and weddings. In addition, contemporary feminist studies only explore the 'burdens' of care-giving without acknowledging the positive experiences of care-receivers. They examine gender inequality and the way care responsibilities devalue the economic status of women (Beckett, 2007). This separation of maternal affection from maternal care may not resonate with how mothers of disabled children approach caregiving or interpret their social experiences of disability (Green, 2007; Ryan & Runswick-Cole, 2009).

Consanguinity

For British South Asian Muslim mothers ableism is enmeshed with racism and xenophobia, in that their children are only valued if they conform to the model minority myth of hardworking high achievers and considered good citizens by the yardstick of Fundamental British Values (FBV). Scrutiny from the government, media and academia has meant that South Asian Muslim mothers have also been problematized for the cultural practice of cousin marriages which has been linked to disability.

It is important to note that consanguineous or cousin marriage does not signify incest; marrying a parent, sibling, grandparent, step parent, or aunts and uncles is prohibited by all mainstream religions including Islam. In the UK, Pakistanis and Bangladeshis have traditionally arranged marriages within extended families either because they already know the other family well, in order to retain wealth within the family, or due to community obligations to the extended family. British Pakistanis consequently have the highest incidences of consanguinity-related disability, mostly between first cousins and second cousins (Darr, 2009).

Due to the prevalence of cousin marriage within the British Muslim community, a cultural rather than religious practice, women's reproductive practices have become sites of scrutiny in terms of their 'efficiency' in expanding and strengthening their diaspora communities. Cousin marriage often disadvantages South Asian women because they not only experience the cultural pressure to produce healthy offspring that maintains the family name, but they also overwhelmingly become full-time carers for their children which halts any personal or career aspirations. Moreover, if their child is diagnosed with a disability, they become hyper-visible and vulnerable to further scrutiny from family, community members, media and broader institutions.

This has been highlighted by the media:

Bradford is 'very inbred': Muslim outrage as professor warns first-cousin marriages increase risk of birth defects. (Kelly, 2011)

Backing for minister over first-cousin marriage comments. (Sparrow, 2008)

This xenophobic narrative positions Muslim mothers as procreating babies with 'birth defects' and placing unnecessary pressure on welfare services. Ironically, the empathy that is afforded to White mothers of disabled children in the media and academia through a 'loss of a perfect child' narrative is missing when it comes to understanding the experiences of British South Asian Muslim mothers. Within academia, there is a perception that South Asian Muslims reject genetic counselling and prenatal testing because of their religious views, however, this is contested by existing research. For instance, Darr et al. (2013), who used focus groups with 50 British Pakistanis, has challenged negative media stereotypes whilst exploring community perspectives about linking consanguinity with disability. They found that despite a tradition of arranged cousin marriages, most participants wanted to learn about the genetic risks of consanguinity. However, there was cultural and community pressure not to discuss sensitive issues which might blame parents for their children's disability. Stigma surrounding disability made it difficult to marry outside one's kin if a family had existing disabilities, which perpetuated marriage within families with existing disabilities; this precluded discussions about genetic testing (Darr et al., 2013). Most participants reported that health information explaining consanguinity and genetic risks often stereotyped consanguinity as a Muslim issue. Contrary to some health studies (Modell et al., 2000), most participants said they would consider termination if pregnancy endangered the mother's life, even though professionals often did not discuss termination assuming that Islam forbids abortion (Darr et al., 2013). Moreover, younger South Asians wanted to choose their own partner, whilst still respecting cultural traditions (Darr et al., 2013).

Shaw and Hurst's (2008) research with 66 British Pakistani Muslims in Southern England is another study which explored attitudes to sharing genetic information within family networks, found that families welcomed genetic counselling but were apprehensive about sharing information with extended families. They also found that men acted as the gatekeepers to information for their families, withholding information either to prevent blame for their child's disability, due to difficulties in explaining medical jargon, or to avoid community pressure to remarry (Shaw & Hurst, 2008). Participants also requested interpreters who could accurately explain the genetic risks to spouses with poor English proficiency. Participants expressed concern that openly discussing possible genetic risks could create family feuds and misunderstandings; most participants wanted information kept between couples to protect their marriages and their children's future marriages from community stigma. Participants also expressed scepticism about the actual risk from consanguinity due to a lack of understanding of genetics (Shaw & Hurst, 2008).

Both studies highlighted the patriarchal traits which affect the willingness of British Pakistanis to tackle consanguinity. These studies also revealed that participants distrusted GPs and the health information about consanguinity's genetic risks, fearing that they were being singled out and scapegoated. There was little evidence suggesting that religion influenced the participants' attitudes to genetic counselling, however, negative stereotypes about Muslims affected the amount of information given to Muslim families, further alienating them and lowering their quality of life.

In my research, whilst five of the eight mothers have cousin marriages, the subject of cousin marriages was not directly approached. However, mothers did broach this topic when discussing support systems, caring responsibilities and their understanding of their child's disability. For instance, Parveen was increasingly aware that many young mothers in her community were unaware or discouraged by family elders to discuss their concerns about their own cousin marriages, and the potential effect it could have on their children and their caring responsibilities.

> I'm bringing this into my sermons. So yes, cousin marriages are allowed but not obligatory, I've informed my siblings and my husband's siblings that my children won't have cousin marriages.—Parveen

Parveen suggested that for British Pakistani diasporic communities to maintain roots with Pakistan through transnational cousin marriages, meant that women were often kept uninformed of the genetic risks whilst simultaneously facing patriarchal pressure to immediately reproduce. There was no discussion by community gatekeepers (mostly men) about the practicalities involved in raising children with the women in the family who were responsible for the care. This trait of deliberately not communicating was passed on from generation to generation, thereby maintaining patriarchy within each family. For instance, Tahira discussed how her son, Farrukh's disability may be attributable to consanguinity and that her husband's learning disability may also have resulted from a lack of intergenerational dialogue on cousin marriage. She was aware that her own marriage might have been arranged for the sole purpose of arranging a life-time carer for her husband. There is a danger that discussing cousin marriages may reinforce, to an outsider's perspective, the narrative that

South Asian cultural practices are oppressive; however, by not including the voices of South Asian mothers, there is a greater danger in ignoring authentic voices on the issue of cousin marriages. Tahira needed practical support from her family, which in her case was absent.

> I feel sad because of my husband, if he would be with me, then I might not have even these problems that I am facing right now … because you cannot compare my spousal support with anyone else.—Tahira

For Tahira, her role as the primary carer for her husband and Farrukh meant she became increasingly aware of the unbalanced and gendered nature of her caregiving responsibilities, and her resulting desperate need for formal care support because informal care support was unavailable.

GENDERED CARE

In recent decades, the UK has experienced a decline in men being their family's sole breadwinner with the gradual increase of two income families. However, the situation is rather different within disabled families where a gendered division of caregiving labour stills exists; this is linked directly to their experiences of poverty. Low-income families with disabled children often struggle to gain access to vital family support, respite and childcare. Consequently, women are often left to do unpaid care work at home; indeed, women constitute the majority of primary carers for disabled children. According to one report, 84 per cent of mothers of disabled children are not in paid employment compared to 39 per cent of mothers of non-disabled children. In fact, only three per cent of mothers of disabled children are in full-time paid employment (Papworth Trust, 2018). When we consider an additional layer of ethnicity and religion, we find that a majority of British Bangladeshi and British Pakistani Muslim women live below the poverty line, bearing the brunt of welfare cuts as well as cultural patriarchy. According to one report, the unemployment rate for British Pakistani/Bangladeshi women was 59 per cent, 33 percentage points difference with White British women (ERSA & PeoplePlus, 2018). High rates of unemployment can also be linked to unpaid responsibilities at home. For instance, a study by the Fawcett Society found that 30 per cent of British Pakistani women and 31 per cent of British Bangladeshi women perform unpaid childcare compared to 6 per cent of White British women (Breach & Li, 2017). Whilst some of this disparity may be attributed to patriarchal elements of South Asian culture, it can also be attributed to an inequitable access and a lack of information about welfare services that allow women from minoritised communities to reduce their unpaid labour.

Academics have explored the experiences of respite care of South Asian disabled families. Shah's (1995) mixed-method study with 35 South Asian families of children with SLD highlighted their low uptake of respite services, which was attributed to parents being unaware of available respite services, or parents presuming that same-sex carers were unavailable; some parents were genuinely concerned that respite care was tantamount to permanently removing their children into social care (Shah, 1995). Despite their unfamiliarity and apprehension towards respite services, many parents still needed formal support but were reluctant to request help believing that their child was their sole responsibility. Shah (1995) suggested that feelings of parental inadequacy could arise from using respite services, which was common to both White British and South Asian families. A decade later, research by Hatton et al. (2004) with 136 South Asian disabled families revealed a similar picture. They concurred with Shah's (1995) contention that the lack of same-sex carers explained why so few South Asian families utilised formal care services. The lack of same-sex carers did help explain the gendered care involved for some of my participants. For instance, whilst Maria did not mind female carers for her son Aamir, but for her youngest daughter she knew her husband would refuse respite care if the carer was male.

> Aamir goes to play schemes and respite ... I don't know if I can leave Saman [her youngest daughter who is also disabled] ... If I leave the girls with people we don't know then ... I don't think my husband would be too happy with that ... I will be the sole carer for her then ... I don't know we just have to take that step when we come to it.—Maria

Although Maria's husband was hands-on at home, he worked long hours as a taxi driver. Most of the physical care for Saman and Aamir was carried out by Maria; this involved changing, feeding, bathing and physiotherapy at home. She reflected that with her non-stop caregiving duties for Saman and Aamir, her other three children had learnt to mother themselves and become young carers for their siblings. Whilst she was able to get some respite care for Aamir, her husband was uncomfortable with Saman being looked after by respite carers. When discussing her husband not consenting to using different sex carers for her daughter, Maria appeared to have prioritised which battles she was willing to expend emotional and physical energy to fight, and which battles she could put off for another day. She recognised that the extent of his practical and emotional support was largely dependent on how his own labour was set out within a capitalist structure. The fact that immigrant communities must often work low paying jobs with long or odd working hours, in effect, becoming less visible within their own family space, ultimately results in physical care becoming gendered. It is entirely possible that in the future, Maria's husband will be less keen on male carers for his daughter, Saman and that would inevitably put the extra burden of care on Maria, however, she would also experience the loss of her spouse being present at home because he is the main breadwinner in a wider capitalist structure that exploits the precarity of his labour. However, Kiran was more forthcoming about gendered care.

> My husband probably wouldn't like me saying this but I feel I'm a single parent because, mind you, he does work long hours. I mean on Mondays he's gone by nine and doesn't come home until half past eight … You know everything is done by then. [laughs] … so I feel I've done everything. You know, Ahmed has had his bath, he's in pyjamas and he's just brushed his teeth and I've just put him to bed. That's about it.—Kiran

Again, labour market conditions are such that Kiran's husband must work long and odd hours, and yet the fact is that her family cannot function on her husband's income alone. Like Maria, Kiran's quality of life and family life had been disrupted, leaving her feeling overworked and exhausted. Kiran was not resentful towards her husband for working long hours, and instead acknowledged the harsh financial realities that not only required him to work such arduous hours but also necessitated that she take up part-time work. Within the current capitalist structure, her family became dysfunctional in that she was acting as a de facto single parent, because she is disproportionately carrying out the physical care as well as contributing financially towards the running of the household. She was performing what Hochschild and Machung (2012) term the 'second shift', one shift at work and the second shift at home; this denotes being active in labour market whilst continuing to perform all or most of the housework and childcare. This is not necessarily an outcome directly borne out of household patriarchy, but rather has trickled down from wider patriarchal capitalist structures that depends on women contributing unpaid physical labour. The same structures ensure that her husband is paid more whilst also insinuating that she would be better placed to be a home-carer, echoing the concerns of bell hooks discussed earlier. Therefore, we cannot simply say that Kiran's husband displays patriarchal behaviour because society at a systemic level has ensured that Kiran, as a woman, must be the one working part-time, being paid less for that work and still performing a disproportionate amount of the physical care at home.

One aspect that literature on the 'second shift' overlooks is how intersectionality affects the choices of mothers like Kiran and Maria. An obvious exclusion is how the 'second shift' affects families caring for disabled children, as well as the lack of recognition of job insecurity when one is from a minoritised community. It is not just that Maria's husband may espouse a patriarchal mindset, but rather that living in the UK as a first-generation immigrant has meant that he has been subject to wider societal structures that exploit and devalue immigrant labour. He is stuck in work/life cycle as a taxi driver, whereby his labour is structured for him in such a way that it reproduces gendered care at home. Kiran and Maria's husbands cannot just quit their jobs even if they wanted to foster more egalitarian values and practices at home. Men and women from minoritised communities have less fluidity and agency to move around in the UK labour market, for fear of facing potentially long periods of unemployment and financial straits because of institutional racism and xenophobia.

> I don't have a family here to help me, I hardly get any [physical] support from my husband, I use the carers whenever I'm at work and when I come back they just leave. I pay them myself [because] I don't get any public funds from the DLA at the moment.—Alina

In Alina's case, the lack of input by her spouse was exacerbated by her temporary legal status on a UK work visa which required her to pay for any formal care she needed for her son, Imran. As she has no recourse to public funds, she was tied into working long hours so she could afford the formal support that Imran needed. From Alina, Maria and Kiran's experiences, it seems that the concept of a second shift and gendered care is more evident when one occupies an outsider status as minorities and immigrants do.

There is also a commonly held perception amongst medical and care professionals that British Pakistani and British Bangladeshi disabled families benefit from an informal support system, such as support from extended families and family friends. However, this perception seems to emanate from literature with able-bodied families from these communities. For instance, Bhatti (1999), Abbas (2004), and Crozier and Davies (2006), all explored the experiences of South Asian non-disabled families within Britain's education system. Although these studies do not examine disability, nonetheless, they do help to contextualise the importance of informal family support systems that are missing from South Asian disabled families such as Alina's.

Studies with South Asian non-disabled families suggest that extended family and community networks play a strong role in their daily lived experiences. For instance, Crozier and Davies (2006) suggest that these strong community ties increase the wealth of social capital and information available to parents, which could be tapped by schools to improve parental participation. Similarly, the US-based study by González et al. (2013) explored how educational practitioners could acquire social and family knowledge from Mexican families to benefit classroom and school practices. Interestingly, Abbas' (2004) mixed-method study suggests that such close community bonds may actually hinder the integration of South Asian families into mainstream society, limiting their social mobility. However, studies with South Asian disabled families reveal an alienated experience; for instance, nearly two-thirds of families in Hatton et al. (2004) reported that they did not receive support from extended family. Katbamna et al. (2004) also challenged the stereotype that South Asian disabled families can avail strong informal networks, resulting in a lower uptake of formal services. Indeed, disabled families report feeling like a minority within a minority, experiencing negative community attitudes towards disability (Katbamna et al., 2004; Hatton et al., 2004). For mothers in this study, this resulted in a double disadvantage: losing out on formal services due to patriarchal structures, and losing out on informal support because of community stigma surrounding disability.

CHILDREN'S SEXUAL RIGHTS

Conversations with children and adolescents about sexual rights and sexuality can be awkward for all parents. It is difficult because it entails negotiating personal boundaries contextualised against a broader background of family values, historical, cultural and religious factors. When these conversations are held at the intersection of disability, the situation becomes further complicated because there are particularly oppressive histories attached to viewing disabled individuals as being asexual and in need of surveillance and protection for their own 'good' (Shakespeare, 2000). Shah (2017) suggests that disabled children and young people are not only excluded from experiencing their sexual rights through different forms of surveillance, but they are also actively prevented from exploring their sexuality compared to non-disabled children and young people. This is underpinned by a deeply infantilising and ableist notion that disabled children are forever *masoom* (pure), incapable and disinterested in exploring their sexual identities. It is reasonable to assume that able-bodied children and adolescents will receive greater agency in the setting of boundaries as well as in exercising their sexual rights, because they have the privilege to control which aspects of their sexuality remain invisible and which aspects they make visible to their parents for their approval. For instance, able-bodied adolescents could regularly attend nightclubs with friends without raising parental suspicions, to some extent making their choices invisible to their parents without being penalised. Rogers (2010) points out that this debate cannot be reduced to maternal rights and values versus the rights of the child. Rogers suggests that such conversations force us to consider who is treated as human, who is afforded dignity, and

> ... about who has the right (and ability) to be sexually and intimately active, to mother as a disabled person and ultimately to make decisions about relationships, intimacy and their own body. What we are also made to consider in such cases is who is governing and is this necessary? (It is often the mother or main 'carer'). (Rogers, 2010, p. 64)

The last point that Rogers (2010) makes above, speaks directly to the extent to which parents and other stakeholders (often medical professionals) can make decisions on behalf of children and young people regarding their reproduction and sexual rights. These decisions often masked under labels such as 'safeguarding', 'protection' and 'genuine care', can impact the choices that disabled children and young people are granted. For instance, the case of a British mother, Alison Thorpe who appeared before the Court of Protection to fight for the right to have a hysterectomy performed on her 15-year-old disabled daughter (Bowcott, 2007). She defended her decision by saying that this procedure would reduce the pain and discomfort of menstruation, stating that, "*I am looking at the interests of an individual, my daughter. I am not suggesting that disabled children as a whole are given this operation.*" It is difficult not to link this explanation to the physical care of her daughter, and the physical labour involved in maintaining her daughter's feminine hygiene during menstruation. In essence, her daughter's dignity is intricately linked to her own labour in performing her physical care. Even if we accept that Alison wanted to maintain her daughter's dignity, the capitalist and ableist system and structures within which she performs her physical labour, makes it impossible for her to recognise and validate her daughter's dignity and individuality. The only way she can resolve her daughter's discomfort and her own labour is by desexualising her daughter. It is an ethical dilemma where we must ask: is governing her daughter's body necessary, and whose sake is it for? Surprisingly, this case was not situated in the broader discourse around eugenics, despite the fact that a disabled young person was having the right to reproduce removed without her consent. In wanting to remove her daughter's womb, she was taking away her daughter's choice and rights to be a mother if she so chose at a later point. This raises the question of who is classed as fully human, with all the rights and freedoms that accompany this label. This issue becomes more contentious if we examine how gender, religion, and more importantly, patriarchal culture mediates this governance and decisions around reproduction and expressing their sexual rights. For instance, in my research, mothers of boys expressed their privilege in not having to deal with a child's menstruation. They could not imagine the hard choices that mothers of girls face to make their mothering more manageable. Maria, mother to Aamir, a 14-year-old boy and Saman, a five-month-old baby girl suggested that it was too early to tell how her mothering would change as Saman grows older.

> I don't know, I haven't been there yet so I can't relate to that, but yes, probably, I would … families I've seen with disabled children when they get older, mothers are still their daughter's carers especially … they've grown into a fully grown adult but they're being looked after. You know these children have fully grown bodies but their mentality is still that of a child, their level of care is the same as for children or babies, totally 100 percent relying on their carer.—Maria

Maria and other mothers of boys reported knowing families who were raising disabled daughters, and they dreaded their daughter's monthly menstrual cycle. Maria found it easier to empathise with mothers as full-time carers having to make those difficult choices, because no one else was performing that level of intimate labour. These mothers also recognised that whilst their child's gender should not affect their views about who gets relatively more freedom to exercise their sexual rights, community and societal patriarchy would ultimately view an adolescent girl's public display of sexuality more harshly than that of a male peer.

Patriarchal culture mediated notions of modesty and the extent to which attitudes differed for boys and girls.

> … if her daughter has no understanding of her sexuality, she will be more vulnerable to sexual abuse, so she [a mother] should be worried … its exactly like if someone would want to steal money from her, so this also very valuable to her … her daughter's privacy, her sexuality and modesty is the child's private property and if the child isn't in a position to protect herself, then her carers/guardians should step in.—Parveen

This discourse surrounding a disabled adolescent exercising their sexual rights is tied to the risk of sexual assault. It is linked to how girls should remain invisible and modest, staying below the radar to avoid any dangerous situation. Obviously, this punishes girls rather than the sexual predators. In mothering able-bodied children, one is able to instil these self-safeguarding techniques to ensure they are not vulnerable to dangerous situations. Parents are able, to some extent, to set boundaries/rules regarding how their children should dress or interact with adults.

Maternal attitudes to their child's sexual rights were ableist. Parveen felt the only way to safeguard her disabled daughter was to desexualise her, knowing that she has considerable 'power' over her daughter to make this decision compared to if her daughter was able-bodied. Parveen's relative power exists because she is situated within an ableist structure that makes it possible for her to choose between preventing sexual abuse, and her daughter's right to her own body. The study by Hussain et al. (2002)[1] with South Asian disabled young people reported that parents placed great importance on protecting a disabled female family member's 'honour' from outside influences, because female honour was perceived as being more easily damaged and less easily repaired than a disabled male

---

[1] Hussain et al. (2002) interviewed 29 disabled young people from South Asian background between the ages of 17 and 30 years old, of which 19 were Muslims and 10 were Sikhs. They also spoke to 14 parents.

individual. They also found that many disabled young females criticised the unequal treatment and lack of socialising opportunities permitted to them by rigid South Asian patriarchal family structures. Mothers in my study did not identify with this notion of preserving female honour in the same way as parents in Hussain et al. (2002). However, a few mothers confirmed that gender did affect how they viewed their child's protection and safeguarding, because girls are more closely scrutinised by the South Asian community than boys are.

Within this research, I also explored how dimensions of a child's sexual and gender rights intersect with religious and cultural beliefs. Mainstream Islam does not explicitly prohibit sex education, however, the religious interpretations of most mothers in the current study disapproved of 'exposing' their child to sex education; this indicates that religion and sexual rights were both enacted within a cultural framework. This was also enmeshed with their religious identities which disapproved of public displays of sexuality and sexual experiences.

> ... you know our family's background is Muslim. Well Daniel was in music therapy when he tried to [sexually] stimulate himself. The music therapy teacher said, 'I don't mind it', but we do ... as parents, we don't like that our son does it, he's lost in his own world and he doesn't know what he's doing.—Maham

Although the dichotomy between parents wanting to control their child's sexual behaviour in public against a child's right to exercise their sexuality is not limited to a Muslim context (Rogers, 2010), nonetheless, it can be reinforced through a religious lens. This chapter also considers the naivety displayed by mothers in my study in dismissing their disabled children as individuals who have sexual needs post-puberty and as such would require sex education, as well as the implications as they transition into adulthood. Although this conservative mindset holds true across all ethnicities, the predominance of religio-cultural values and beliefs within South Asian communities which recoils from such conversations prevents disabled people from being viewed sexual beings.

> ... they're going to teach them about sex education and I laughed. How are you going to teach them? What will you tell them? Because of our culture and religion, their [White people's] norm and our norm are very different. We don't explain half the things to our children that White people do, they

probably know from the very early age but we keep it hidden from our children for as long as we can ... girlfriends and boyfriends, gays and lesbians, it's not allowed in our religion, we teach them that. It's not the norm and I have to respect that.—Maria

Both Maham and Maria reflected how opportunities for their disabled child to explore their sexuality through sex education and masturbation were forbidden by both their South Asian culture and their Muslim faith; indeed, mothers felt that sex education risked jeopardising their 'Muslim' values. Their religious interpretations were deeply entwined with what disabled boys and girls could or could not do, based on whether normative culture deemed it acceptable. Whilst it was unacceptable to Maham that Daniel was permitted by teachers to masturbate in school *in front* of others, her disapproval was based on its prohibition in Islam. The issue was not that her son was masturbating in school, but rather that he was masturbating at all which is deemed a sin before marriage according to Islamic teaching. It is entirely plausible that many able-bodied Muslim young people are able to hide their sexual practices, including masturbation, from their parents and escape any type of reprimand and atonement for breaking this implicit religious rule; however, for disabled Muslim young people this privilege may not be available, resulting in them being chastised and disciplined more often for their sexual practices. The maternal responses in my study regarding their children's sexual rights also echoed Shakespeare's (2000) claim that disabled people are often perceived as being either asexual, or prone to inappropriate sexual behaviour. Mothers were naïve in denying their disabled children's post-pubescent sexual needs, and who consequently would require sex education.

This bias holds true across all ethnicities. Although there is a dearth of literature examining South Asian parental attitudes towards the sexual rights of their disabled children, a quantitative study by Isler et al. (2009)[2] on the attitudes of Turkish parents of disabled children reveals that three-quarters of parents did not have any sex education training, while over 32 per cent of parents had never discussed sex with their disabled children. Moreover, contrary to the mothers in my sample, most parents wanted sex education to be a part of their child's school curriculum (Isler et al., 2009);

[2] Isler et al. (2009) involved questionnaires with forty parents who have children with mild to moderate difficulties with age 15 and older and who attended an occupational school in Turkey.

this may be considered surprising since this study was conducted in Turkey, a Muslim-majority country. The findings from Isler et al. (2009) led me to postulate that perhaps attitudes to sexual rights were not merely shaped by religion, but that culture also had a dominant influence.

This site of conflict with regard to parental authority that regulates their disabled child's sexual behaviour in public spaces, versus that child's right to exercise their sexuality is not only found within the Muslim community. Rogers (2010), in her research with 24 parents of disabled children, found that 4 parents reported that their child had displayed inappropriate sexual behaviour in public. Parents in her research framed this issue from a safeguarding viewpoint. Whilst it is difficult to ascertain why parents gatekeep and set boundaries on their children's sexual behaviour and sexuality, the consequences of such decisions are real and can stifle the individual rights of their children. Rogers (2010) posits that it would be unfair to assume that all parents who are worried about safeguarding their child from vulnerable situations are in effect disabling that young person's right to assert their sexuality. Moreover, she suggests that this topic requires further debate specifically around mothering, rights to sexual intimacy, reproduction and intellectual disability.

A common concern amongst all mothers in my study was safeguarding their child against the risk of sexual abuse. As discussed previously, following extensive media reporting about the sexual abuse of children, mothers became concerned that protecting their disabled child was a significant challenge because he/she may not be able to recognise sexually abusive behaviour from others. This affected their willingness to send their children on overnight school trips or short breaks. This suggests that this issue transcends the need to recruit same-sex carers to increase the uptake of formal services, contrary to the findings of the mixed-method study with 54 South Asian disabled families conducted by Hatton et al. (1998). Indeed, some mothers expressed hesitance to send their child away for longer periods even with same-sex carers.

> ... they should understand the child is special, you shouldn't even send a normal child away like that ... explain that it's better if I accompany her.—Maham

Maham's concerns reflect broader South Asian attitudes towards permitting their children on overnight school trips, which is a common practice within British schools. Some mothers admitted that they were equally uncomfortable with allowing their non-disabled children to stay overnight away from home as they were with their disabled child. They recalled events from their own lives growing-up and how they were never allowed sleepovers or overnight school trips.

Like Maria, Shehnaz also had a son and a daughter with disability; however, she did not require respite support because her older non-disabled children helped her with domestic and caring responsibilities. Consequently, it was difficult to ascertain with any assuredness whether Shehnaz would use respite services for both her son and daughter, or whether she would adopt a gendered lens. This was also a significant concern when I introduced vignettes[3] to the mothers in my study, because oftentimes it was difficult to ascertain whether they were relating their personal experiences or discussing the fictional characters from the vignettes. Most decisions about the use of same-sex carers were mediated by the willingness and stance of spouses. Interestingly, Parveen and Tahira, the remaining mothers in my study who did utilise respite services and both of whom were first-generation immigrants, both also had British-born spouses who had experience of the UK education system; however, Maria's husband was a Pakistani-born British citizen who had no British educational experiences. Although this study is small-scale, discussions with mothers show that Pakistani fathers may influence their families' openness to using formal services.

EDUCATION, EMPLOYMENT AND MARRIAGE

Mothers overwhelmingly reported that they desired equal access to educational and employment opportunities for their sons and daughters. However, notably, a few mothers did highlight how their school placement decisions had been unduly influenced by male members of their extended families. Notwithstanding this patriarchy, these mothers acted as *"agents of change"* (Bhatti, 1999, p. 86) within their families, because they challenged traditional family decisions and patriarchal decision-makers. For instance, mothers countered gendered expectations by rejecting the idea of an arranged marriage for their disabled children purely as a care provision in their adulthood, expressing higher expectations of government support in their children's post-19 and post-25 years. This suggests

---

[3] I used fictional stories written in the third person to discuss sensitive topics around gender, religion, culture and immigration.

that mothers wanted their sons and daughters to be independent adults in their own right, rather than to have arranged marriages as a way to address their long-term caring needs.

In her ethnographic research with non-disabled British South Asian families, Bhatti (1999)[4] found that whilst girls advanced onto higher education and employment their rationale was not to attain financial independence. Bhatti found that South Asian girls sought employment to either earn money to send to or visit their country of origin, to earn money for their dowry, or to provide evidence to the Home Office that they could financially support their would-be husband from their country of origin. However, mothers in my study expressed their disapproval of marriage for their disabled child in the future.

> I know Ahmed could never get married because he can never take responsibility for himself, so how can he take responsibility for his wife? If she's [fictional mother in the vignette] thinking about someone who'll take care of her daughter, he might just want to come over to get British citizenship, well that's what most people do, don't they? And once he's here, he might say, 'ok I'm off'. So there's no guarantee that he'd look-after her.—Kiran

Kiran's views largely echoed the sentiments of the other mothers in the current study. Whilst marriage is perceived as a priority within South Asian culture, mothers were worried about the potential abuse that their children, specifically their daughters could be exposed to if potential spouses were offering marriage only because they were interested in gaining UK citizenship. Their concerns raise several points of analysis; firstly, mothers such as Kiran conceptualised marriage as far more than merely a carer's role, and felt that a one-sided relationship would be unfair where one partner solely bore the burden of the relationship. This suggests a possible divergence in how mothers perceived the role of their disabled child's partner compared to the roles within traditional South Asian marriages. This more progressive attitude may be the result of the mothers' own experiences of transnational marriage, and their observations of marriage within the British South Asian community more generally.

Secondly, mothers in my study were opposed to transnational marriage. This runs contrary to Bhatti's (1999) suggestion about South Asian

---

[4] Bhatti (1999) in her research engaged with 50 Asian families from Pakistani, Bangladeshi and Indian backgrounds, of whom majority were Muslim families and her research also included interviews with 50 boys and girls.

parents striving to maintain links with their home country by arranging transnational marriages for their children, and in the process coming closer to what she termed the "*myth of return*" (p. 9). Whilst the mothers recognised that marriage was a priority within South Asian culture, they were worried about the potential abuse that their children, specifically their daughters, could be exposed to. They feared that any potential partners would only agree to marry their disabled child to gain legal entry into the UK, and once they had gained UK citizenship, they would leave their child. The notion of arranging transnational marriages for disabled children has also been explored by Hussain et al. (2002). Parents perceived that offering a potential spouse the chance to legally settle in the UK was a win-win situation because this could secure a permanent full-time caregiver for their child. Over a decade after Hussain et al. (2002), mothers in my study rejected this practice. This may have been because there are now stricter immigration controls in place in the UK, or because of the dangers of marriage fraud which could harm their child, or the fact that the new SEND Code of Practice now provides support until the age of 25 years, or all of the above. Mothers were aware that marriage was not the only future option for their child, particularly because the government offers care assistance, as they grow older.

Thirdly, mothers did not want their child to be perceived as a burden by a potential spouse. Although they recognised that caregiving was a significant aspect of their child's disability, they also perceived that there was far more to their child's personality. Moreover, such an arranged transnational marriage would disregard their child's happiness and well-being. Maham suggested that she would only consider marriage for her son, Daniel if he developed an emotional attachment to someone, like a classmate.

Finally, all mothers stressed the importance of providing their child with valuable life skills or education to improve their independence. This suggests that mothers in my study held greater expectations that professional services and educational institutions would be able to make their children more independent, and consequently, less reliant on family networks to care for them as they grew up. Despite their views on marriage for their disabled children, mothers were aware that South Asian culture holds a different standard for boys and girls, with greater pressure on girls to get married early, which itself was 'disabling' for all South Asian girls.

In relation to educational opportunities, mothers generally agreed that there should be no difference between boys and girls.

I think everyone should have an equal chance, girl or boy, but only if the child shows an interest in developing their education. But if they're not showing any signs of that, they have to roll onto other options.—Shehnaz

Mothers were aware that their disabled children had the option to attend vocational colleges, or enrol in various life skills courses, or take on apprenticeships if either they were not interested or did not have the aptitude to continue in higher education. Contrary to Bhatti's (1999) study, where girls felt ambivalent about their future education because of gendered expectations regarding early marriage, mothers in my study discussed the future educational opportunities for their daughters in a less constrained environment. This was possibly because there was little or no maternal focus on marriage which can often obstruct educational aspirations, and education was viewed as a tool to make their daughters more independent and capable of work that was suited to their needs. All mothers with daughters aspired to give them the opportunities that they had not had themselves growing up. This was most evident when Saira and Shehnaz both revealed how their elder daughters were now attending further or higher education institutions, and that whilst marriage was an important part of life, it should not compromise their daughters' opportunities for self-development.

References

Abbas, T. (2004). *The education of British south Asians: Ethnicity, capital and class structure*. Palgrave Macmillan.

Beckett, C. (2007). Women, disability, care: Good neighbours or uneasy bedfellows? *Critical Social Policy, 27*(3), 360–380.

bell hooks. (2000). *Feminist theory: From margin to center*. Pluto Press.

Bhatti, G. (1999). *Asian children at home and at school: An ethnographic study*. Routledge.

Bowcott, O. (2007, October 8). Mother defends hysterectomy for disabled daughter. *The Guardian*. Retrieved December 15, 2020, from http://www.guardian.co.uk/news/2007/oct/08/medicineandhealth.uknews

Breach, A., & Li, Y. (2017). *Gender pay gap by ethnicity in Britain—Briefing*. Retrieved December 15, 2020, from https://www.fawcettsociety.org.uk/gender-pay-by-ethnicity-britain

Bukhari, M. I. I. (1966). *Sahih Bukhari*. Muhammad Sarid.

Cheruvallil-Contractor, S. (2016). Motherhood as constructed by us: Muslim Women's negotiations from a space that is their own. *Religion and Gender, 6*(1), 9–28.

Crozier, G., & Davies, J. (2006). Family matters: A discussion of the Bangladeshi and Pakistani extended family and community in supporting the children's education. *The Sociological Review, 54*(4), 678–695.

Darr, A. (2009). Cousin marriage, culture blaming and equity in service delivery. *Diversity in Health and Care, 6*(1), 7–9.

Darr, A., Small, N., Ahmad, W. I., Atkin, K., Corry, P., Benson, J., & Modell, B. (2013). Examining the family-centred approach to genetic testing and counselling among UK Pakistanis: A community perspective. *Journal of Community Genetics, 4*(1), 49–57.

ERSA & PeoplePlus. (2018). *Race, ethnicity and employment: Addressing disparities and supporting communities.* Employment Related Services Association (ERSA). Retrieved December 15, 2020, from https://ersa.org.uk/documents/race-ethnicity-and-employment-addressing-disparities-and-supporting-communities

Gabel, S. L. (2018). Shatter not the branches of the tree of anger: Mothering, affect, and disability. *Hypatia, 33*(3), 553–568.

González, N., Moll, L. C., & Amanti, C. (2013). *Funds of knowledge: Theorizing practices in households, communities, and classrooms.* Routledge.

Green, S. E. (2007). "We're tired, not sad": Benefits and burdens of mothering a child with a disability. *Social Science & Medicine, 64*(1), 150–163.

Hatton, C., Akram, Y., Shah, R., Robertson, J., & Emerson, E. (2004). *Supporting south Asian families with a child with severe disabilities.* Jessica Kingsley Publishers.

Hatton, C., Azmi, S., Caine, A., & Emerson, E. (1998). Informal carers of adolescents and adults with learning difficulties from the south Asian communities: Family circumstances, service support and carer stress. *British Journal of Social Work, 28*(6), 821–837.

Hochschild, A., & Machung, A. (2012). *The second shift: Working families and the revolution at home.* Penguin Books.

Hussain, Y., Atkin, K., & Ahmad, W. (2002). *South Asian disabled young people and their families.* Joseph Rowntree Foundation, The Policy Press.

Isler, A., Beytut, D., Tas, F., & Conk, Z. (2009). A study on sexuality with the parents of adolescents with intellectual disability. *Sexuality and Disability, 27*(4), 229.

Katbamna, S., Ahmad, W., Bhakta, P., Baker, R., & Parker, G. (2004). Do they look after their own? Informal support for south Asian carers. *Health and Social Care in the Community, 12*(5), 398–406.

Kelly, T. (2011, May 30). 'Bradford is very inbred': Muslim outrage as professor warns first-cousin marriages increase risk of birth defects. *Daily Mail*. Retrieved December 15, 2020, from http://www.dailymail.co.uk/news/article-1392217/Muslim-outrage-professor-Steve-Jones-warns-inbreeding-risks.html

Modell, B., Harris, R., Lane, B., Khan, M., Darlison, M., Petrou, M., & Varnavides, L. (2000). Informed choice in genetic screening for thalassaemia during pregnancy: Audit from a national confidential inquiry. *BMJ, 320*(7231), 337–341.

Papworth Trust. (2018). *Facts and figures 2018: Disability in the United Kingdom*. Papworth Trust. Retrieved December 15, 2020, from https://www.papworth-trust.org.uk/about-us/publications/papworth-trust-disability-facts-and-figures-2018.pdf

Rogers, C. (2010). But it's not all about the sex: Mothering, normalisation and young learning disabled people. *Disability & Society, 25*(1), 63–74.

Runswick-Cole, K., & Ryan, S. (2019). Liminal still? Unmothering disabled children. *Disability & Society, 34*(7–8), 1125–1139.

Ryan, S., & Runswick-Cole, K. (2009). From advocate to activist? Mapping the experiences of mothers of children on the autism spectrum. *Journal of Applied Research in Intellectual Disabilities, 22*(1), 43–53.

Shah, R. (1995). *The silent minority-children with disabilities in Asian families*. National Children's Bureau.

Shah, S. (2017). "Disabled people are sexual citizens too": Supporting sexual identity, well-being, and safety for disabled young people. *Frontiers in Education*. https://doi.org/10.3389/feduc.2017.00046

Shakespeare, T. (2000). Disabled sexuality: Toward rights and recognition. *Sexuality and Disability, 18*(3), 159–166.

Shaw, A., & Hurst, J. A. (2008). "What is this genetics, anyway?" Understandings of genetics, illness causality and inheritance among British Pakistani users of genetic services. *Journal of Genetic Counselling, 17*(4), 373–383.

Sparrow, A. (2008, February 11). Backing for minister over first-cousin marriage comments. *The Guardian*. Retrieved December 15, 2020, from https://www.theguardian.com/politics/2008/feb/11/religion.medicalscience

The Hadith (Sunan an-Nasa'i 3104, Book 25, Hadith Number 20).

CHAPTER 5

Mothering in Cultural Bubbles

How is South Asian Muslim culture perceived in this country? Is it similar to Alice's perception in Chap. 1, who presumed that all Bangladeshis have large families and live in crowded homes within joint family systems, where women do not have freedom to 'go out' or 'speak out', and where their arranged marriages lack any romance? Such perceptions of South Asian culture are assumed to originate from Islamic dogma and as such be set in stone. Mahmood Mamdani in his book, *Good Muslim, Bad Muslim* (2005) refers to the problematic ways in which Muslim culture is theorised in the West as "*Culture Talk*" (p. 17). He points to the static and dangerous representation of Muslim culture post-9/11 in the USA, and the manner in which that has positioned American Muslims as incompatible and non-integrationist to living in the West.

> Culture Talk assumes that every culture has a tangible essence that defines it and it then explains politics as a consequence of that essence … Culture Talk focuses on Islam and Muslims who presumably made culture only at the beginning of creation, as some extraordinary, prophetic act. After that, it seems Muslims just conformed to culture. (Mamdani, 2005, p. 17/18)

Mamdani's words have stayed with me as I have grappled with the task of discussing how mothers in this study have discussed culture. Should British Pakistani culture be seen as something that is holding mothers back from availing the most that the special education needs system has to

© The Author(s), under exclusive license to Springer Nature Switzerland AG 2021
S. Rizvi, *Undoing Whiteness in Disability Studies*,
https://doi.org/10.1007/978-3-030-79573-3_5

offer? What is British Pakistani culture? Does it acknowledge its blurred boundaries with religion and diverse immigration trajectories? A starting point would be to acknowledge that there is no static or authentic representation of British Pakistani culture or the Islam that is practiced in the UK. Many researchers acknowledge that their research participants may be articulating their experiences in binaries when discussing their diasporic identity and cultural clashes in a South Asian context, dismissing the roles of other mediating actors. Bolognani (2009) has termed this the 'religion versus culture' paradigm; this presents the view that any discourse on the survival of South Asian cultural traditions and practices in the UK is separate and sometimes in conflict with arriving at an "*allegedly acultural, pure and universal Muslim mode of practice derived from the religious texts*" (Bolognani & Mellor, 2012, p. 212). Bolognani (2009) argues that most scholarly work with British South Asian women highlights how women engage deeply with an authentic understanding of Islam, using textual-based interpretations to tackle any oppressive cultural-based practices that seek to undermine their agency in various aspects of their life. She argues that whilst this type of narrative may seem liberating and authentic, nonetheless, it poses two problems. The first issue is directly relevant to understanding how British Pakistani mothers of disabled children can exercise agency to better the experiences of their children. That is, Bolognani (2009) suggests that no evidence-based research has established whether British South Asian women have successfully used an authentic understanding of religious texts to address patriarchal customs and traditions, to enable them to live their lives on their own terms.

> Yet, while these young women were certainly arguing for their God-given rights as Muslim women, as Jasminah's experience of the rishta [marriage proposal] indicates, it was not always clear how these demands were being heard or how family and community responded to these challenges. (Bolognani & Mellor, 2012, p. 217)

Bolognani and Mellor (2012) argue that the efficacy of deploying the religion versus culture paradigm is unclear; rather it illustrates the tensions in negotiating one's diasporic identity. From this perspective, it could be argued that Parveen's grounding in 'proper Islam' has not really changed her daughter's experiences of inclusion and exclusion within SEN provisions—a point which I will revisit later. Secondly, they argue that this discourse is acritical of the coloniser/colonised relationship in Britain where British Muslims, as a consequence of state discourse on Islam and terrorism, feel the need to defend Islam whilst also feeling a sense of inferiority as far as their individual and collective identity is concerned. This coloniser gaze has resulted in British Muslims vehemently opposing the cultural traditions of their older generations as well as aspects of their host British culture, instead favouring what they perceive as a stronger and more authentic religious identity as Muslims. Bolognani and Mellor (2012) posit that this return to 'proper Islam' by British Muslims is not authentic or is devoid of any cultural influences. Given that the majority of Muslims living in Britain belong to the Sunni sect of Islam, they argue that this so called authentic Muslim identity is in fact a Saudi-based cultural understanding of Islam post 9/11. It is not an authentic response because the emergence of this Muslim identity has not occurred in a cultural void. It has been formulated in the context of the 'Saudisation' of Islam in the UK. In attempt to detach themselves from their older generation's South Asian traditions, the new generation has ended up adopting the cultural understanding of Islam as understood and propagated by Saudi Arabia throughout the Sunni world.

The concerns raised by Bolognani and Mellor (2012) have serious implications, and compel scholars researching in the British Pakistani context to be cautious about how they analyse their participants' narratives. However, this caution should not prevent us, as insiders, from researching and presenting such narratives. In line with feminist traditions, it is almost antithetical to feminist goals to argue whether or not mothers in this study exercise false consciousness when they state that they feel empowered by their religious knowledge and understanding, even if that religious knowledge has been formed against the backdrop of Saudi culture. Furthermore, as highlighted in Chap. 4, mothers like Parveen were not only formulating their own religious understandings around care and responsibilities towards their disabled children, but they were also critical of Saudi cultural influences on religiosity and how this could worsen their caring experiences if they focused excessively on what was frowned upon from a cultural standpoint (e.g. the use of opposite sex carers, using overnight respite services, etc.).

> Linked to the fear of stigma of having intellectual disability is the cultural norm of 'shame' attached to accessing health and social care provisions … South Asian's cultural desire for privacy is demonstrated in the ways they approach (and avoid) health professionals … it is perhaps unsurprising that in South Asian communities, life tends to revolve around relationships within the family. (Bhardwaj et al., 2018, p. 255)

The paper by Bhardwaj et al. (2018) is interesting not only because it delves into the myths of shame that create a fixed and archaic image of South Asian culture in the minds of a Western audience, but more importantly because it addresses the role of institutional racism as a separate and parallel story. The study by Bhardwaj et al. (2018), like many studies that examine health and social care access, presents cultural and language differences as barriers that may explain the low uptake of formal services by South Asian disabled families. In most research studies examining access to care for disabled families in the British South Asian community (Azmi et al., 1997; Hatton et al., 2004; Bhardwaj et al., 2018), institutional racism is rarely acknowledged as being tied to their daily lived experiences, and yet institutional racism profoundly determines how these communities internalise beliefs about their capacity and ability to look after and advocate for their disabled family members as well as access to relevant linguistic and cultural tools. What is most consistently projected by a majority of these studies, is the overarching negative cultural stereotype that disability is attached to notions of family shame and stigma which in turn may help explain why these communities are hesitant to seek access. Interestingly, my previous research (Rizvi, 2015) suggests that mothers themselves do not hold negative perceptions of disability nor view disability as something to be ashamed of; nonetheless, they are aware that cultural shame or stigma about disability may be prevalent in their local British Pakistani non-disabled community, and as a result they are more likely to break their ties with such community members. This also mirrors the small-scale research by Akbar and Woods (2020) with British Pakistani mothers, that revealed that families were stigmatised and often forced to remain invisible by extended families and the wider British Pakistani community. Although they acknowledge the absence of negative perceptions of disability among mothers in their study, nonetheless, they suggest that professionals should 'educate' or 'correct' parental perceptions of disability where necessary, just in case they use derogatory terms such as "*pagal (mad), mental or a donkey*" (p. 14) for their own disabled children. This tendency to acknowledge that parents do not hold negative cultural perceptions of disability whilst still mentioning it as a common parental perception within South Asian disabled families is not just specific to Akbar and Woods (2020), and perhaps serves as a pretext to different research studies looking into the experiences of disabled families. However, this framing can be problematic because it presents South Asian families as being dysfunctional, and attributes their poor experiences to being opposed or unfamiliar with the Western conceptualisation of inclusion, disability and the social model of disability, rather than understanding that South Asian disabled families experience exclusion from formal and informal support systems and that the negative perceptions of disability may exist outside of the individual family unit as part of wider community attitudes.

With this current research, it became quite visible early on that mothers discussed culture in two ways: firstly, how they had forged new cultural identities that helped them to better navigate their child's disability. Secondly, they discussed culture as represented within communal and institutional spaces.

CULTURAL BUBBLES

Many mothers in my research lived in neighbourhoods with high South Asian populations and had easy access to mosques, halal butchers, ethnic supermarkets, Pakistani restaurants and takeaways and Pakistani beauty parlours akin to the rich ethnic neighbourhoods described by Bhatti (1999). This close proximity afforded them convenience in terms of their grocery shopping or eating out, but invited increased scrutiny by local South Asian community members. Crozier and Davies (2008)[1] have posited the notion of British South Asians being a *"gemeinschaft–like community, a traditional folk relationship based on locale, kinship and friendship commitments"* (p. 680), one that supports the division of labour within families and positively affects a child's schooling experiences. In contrast, my research found that although the mothers in my study predominantly resided in ethnic neighbourhoods, their families functioned as single distinct units rather than as part of the wider South Asian community, from which they often felt marginalised and estranged. Mothers recalled how unsupportive some community members had been towards their family situation, and how they often avoided interaction with these community members because of the risk of their private family matters becoming fodder for local community gossipmongers. Furthermore, unlike the mothers in Bhatti's (1999) study who felt concerned and insecure about passing down their cultural values to their children, mothers in this current study

[1] In their study, Crozier and Davies included 157 non-disabled families and 400 family members belonging to British Pakistani and British Bangladeshi backgrounds to explore family perspectives on children's educational experiences.

had formulated their own cultural identities and values that may or may not have mirrored those of the British Pakistani community at large.

Bhatti's study (1999) is instrumental in that it offers rich insights into the concerns of British South Asian parents, and their concern that they might 'lose' their children to the values of Western society. Interestingly, mothers in the current study seldom expressed fear about their child possibly losing their cultural identity to Western values; this might be explained by the fact that they had experienced exclusion by the very community with whom they shared their collective cultural identity, as a result of their child's disability. In some instances, this exclusion led the mothers to create their own small 'cultural bubbles' which allowed them to feel a sense of belongingness to an ethnic-Pakistani group, without feeling judged or scrutinised in return. I define a cultural bubble as a small but meaningful space which represents people and cultural symbols that most closely reflect one's inter-categorical positioning and values. It does not just refer to cultural similarity, but rather a cultural bubble entails those people with whom one aligns in terms of religiosity, educational background, family values, language, and so on. Individuals can occupy and belong to different cultural bubbles at the same time, without necessarily fragmenting themselves. I find that this notion of a cultural bubble is an indicative example of how the British Pakistani disabled diaspora is currently navigating and creating their own cultural identities and values based on their own inter-categorical positioning. This detachment from previous notions of a traditional South Asian cultural identity may not be specific to disabled families, nor is it tied to a disabled identity. For instance, Chanda-Gool's (2006) research with South Asian communities in the UK revealed that, rather than viewing this as a generational issue where second and third generation British South Asians have "*lost*" their culture and adopted a Western worldview, one should view this as a dynamic process which highlights how younger generations have developed better "*bicultural skills*" (p. 103). She adds,

> It is characterised by creativity, imagination and the courage to challenge accepted beliefs and practices. It blends a capacity to create a sense of self and retain a sense of belonging without compromising independence of mind and initiative ... This pursuit of alternative self-definition, which synthesises contrasting cultural beliefs and values, is essential to the survival of South Asian identity both in the UK and at global level. (Chanda-Gool, 2006, p. 103/105)

Chanda-Gool (2006) posits that this new imagining of cultural identity is not free from tensions and challenges, and that South Asian individuals face having to decide which aspects of their culture they wish to keep and which aspects they can discard. Participants in her study demonstrated the contradictory and rather messy lives of the diasporic community; they wanted to question and critique aspects of their culture and yet they showed an awareness that this culture was also dynamically transforming, and that a new understanding of their cultural identity and values were always being created. However, this new imagining is being created against the backdrop of a constant institutional gaze which focuses on their cultural deficits rather than their cultural strengths (Harry, 2002). For instance, academics and practitioners have attributed the high incidence of disability amongst the British South Asian community to the cultural practice of first-cousin marriage; however, professionals often overlook the maternal agency that exists within marital practices and the differing maternal experiences that develop out of divergent ethno-cultural backgrounds and migration trajectories. As a result, professional expertise is developed from a one-sided stereotype where Eastern cultures are viewed by professionals as patriarchal, oppressive and inferior to the 'progressive Western' cultural paradigm. Over time, these principally inaccurate and imprecise stereotypes become accepted as 'fact', misinterpreting minority experiences and ultimately misdirecting provisions ostensibly aimed at them.

For all mothers in the current study, this new cultural understanding was deeply entwined with their child's disability.

> I think we've gone passed that stage. I think if he'd been okay, if he'd been normal then I think we would've had a problem. I think we've become more broad minded and more, in a sense, open to things. So culture or however, you know, if we're British or we're Punjabi it's become, it's just not a priority anymore. I think since Daniel has come into our lives we're not really bothered.—Maham

> You know, we are sort of born and bred here … that bit as well. But I don't want to lose my cultural identity as well as my roots … my mum and dad were from back home, from Pakistan you know … I want to keep that side of it as well. Yes, we're British but you know, we don't do all the British things … we don't go to pubs … we're not going to be drinking alcohol and we're not going to have girlfriends or boyfriends … you know, that's all sort of the West … very Westernised aren't they? I want to keep my identity as

well ... well, obviously with Ahmed, you know he isn't going to be able to go out and be able to ... with Ahmed it's a bit different isn't it? He isn't that able 15-year-old boy who is going to be going out by himself, is he? So he isn't going to be able to make those choices.—Kiran

I don't think about it [culture], not for Aamir anyway. I think he's excused from everything because he's got no knowledge, he's got no speech, his mentality is that of a twelve-month-old baby ... Cultural background, when it comes to that I don't think I'm too much into our sort of culture. I think I'm more [attuned to] British culture as much as the next person. I know English so I don't struggle with understanding them and I can put my own views forward. I don't think I've clashed in that respect, even as far as religion is concerned.—Maria

Just as a child's disability has become an important consideration when examining the gendered and sexual experiences of children and young people, so too has it become a significant factor in how culture is being experienced by disabled children and young people. This does not, however, constitute a complete disregard for cultural values and ethos but rather, as Chanda-Gool (2006) suggests, a slow and gradual process of choosing which aspects of their culture they wish to keep and which to disown. For instance, whilst Maham claimed that she was no longer tied to cultural norms, at other instances she stated how she was privileged to live in a joint family system where care could be shared amongst family members and where she could benefit from her mother and sister-in-law's advice on rearing Daniel. It is important to note that Maham construed and valued cultural identity and values quite differently for Daniel compared to her two other children who were able-bodied. With Daniel, cultural identity was strongly enmeshed with a religious identity (such as a focus on eating only halal food, or actively preventing him from masturbating), rather than about non-religious traditions such as an insistence on wearing Pakistani clothes, eating Pakistani food, or speaking Punjabi and Urdu at home. It could be argued that for able-bodied children and young people, such non-religious traditions become a gatekeeping exercise to the British Pakistani community in that as they reach a marriageable age, both young men and women need to prove to the community that they belong to 'cultured' families which still embody traditional South Asian family values. However, this is not to say that all South Asian able-bodied children adhere to, or indeed are forced to adhere to a more static representation of cultural values; as I have previously argued, cultural identity and values are not static and are themselves constantly changing. In addition, the effect of double consciousness on intergenerational diasporic families means that inevitably they do not inhibit binary cultural identities—either British or Pakistani—rather that they have to constantly negotiate their traditions and customs against the backdrop of an ever-changing macro environment.

With disabled children and young people, this expectation to prove their 'cultured-ness' is taken on by their parents. It is mothers who decide the terms of their disabled children's cultural engagement and in doing so, they are actively creating their own cultural bubbles. For instance, Kiran suggests that she would not need to worry about Ahmed going for a drink in the pub or having girlfriends because as she suggests, *"with Ahmed it's a bit different isn't it?"*. She states that she faces a different set of cultural expectations that are entwined with her religious practices. For instance, she revealed how Ahmed's schools had had no understanding or appreciation of the Islamic concepts of *najasat* (impurity) and *pakeezgi* (purity)[2] when training him for toileting, and that she had explained to his care assistants that it was not okay with her from a hygiene or religious standpoint that Ahmed was permitted to wear soiled underpants at school. She also suggested that since Ahmed was entirely dependent on her for his personal and wider care, the likelihood was negligible that he would ever independently go and explore those Western practices that were in direct contravention of his religious faith, such as hanging out in the pub or having sexual relations outside of marriage. There is also a danger of misreading Kiran's narrative—*"very Westernised aren't they?"*—as a case of 'us versus them'. Although she clearly differentiates between Western and Pakistani cultural practices in the context of what is religiously permitted, nonetheless, she is also actively developing her own bicultural identity that takes aspects of both cultures. As a mother, she is acutely invested in holding onto the ethno-cultural values that she was raised with, but not as a mere passive recipient. Kiran wants Ahmed to embrace his British Pakistani identity, to know that he has family in the North of England that celebrates Muslim festivals and Pakistani customs, and that he is part of them.

Of course, there were 'Western' activities that Kiran, Maria and Maham were happy for their sons being a part of, such as participating in school

---

[2] State of purity and impurity—In Islam, after an individual has passed urine or faeces, they need to wash themselves with water to be considered *pak* (pure or clean).

Christmas plays or being part of musical events. It would be too easy to declare that these maternal responses were undermining their disabled child's agency in developing their own individual cultural identities—and to leave it at that. However, there are consequences if mothers do decide to take a step back from this responsibility, as community elders and extended family members are likely to fill the cultural void left by mothers to step forward as purveyors of cultural policing. These individuals are not involved in the day-to-day care of the disabled child or young person, yet still hold strong views about how children should be raised within a certain ethno-cultural identity, one that may not sit well with the family's cultural values or their caring capabilities. In my conversations with Maria and Shehnaz, they revealed that they have had to resort to acting as a guard or barricade against community and extended family members who insisted on offering their opinions and judgements about how they should raise their disabled children. For example, they reported receiving suggestions about teaching their disabled child the rather antiquated Pakistani social etiquette of how to greet elders which may have caused the child added anxiety, or wearing fancy Pakistani clothes which may have affected how quickly mothers could change their child for toileting, or attending cultural events within venues that were not disabled friendly; it is notable that none of these cultural practices would have added anything of value to either the cared for child or the mother. So whilst there is an explicit acknowledgement from mothers that they actively determined which aspects of culture they wanted to keep for their disabled child, the analysis must go beyond the scope of an ableist lens to one that incorporates an intersectional perspective. If dominant cultural practices within the British Pakistani community are to become inclusive, they need to consider the logistics of gendered care, the ways in which certain cultural practices are ableist and which exclude participation by the disabled community, and a consideration of how these practices can create a stronger sense of communal care for disabled children and young people similar to the shared child-rearing practices that bell hooks (2000) suggests.

Notably, I also found that Maria had internalised certain fixed notions of what British cultural values meant to her. Her responses project an acknowledgement and internalisation of the stereotypical notions that British mainstream society, media and the government hold about religious Muslims—that they are not only conservative but also separatist in the sense that they do not have good English proficiency. Whilst both Kiran and Maria were born in this country, Maria felt the need to assert her Britishness; as Bolognani and Mellor (2012) caution us, it appears on the surface that her affirmation of Britishness may have been at the expense of her Pakistani cultural identity. It could be argued that this defence is a natural consequence of her double consciousness, and her deep awareness that her hijab and other patently religious Islamic clothing are viewed as symbols of her lack of integration by the dominant constituents within British society. She felt that she needed to state explicitly that she was just as British as the next person was and that there was no issue of cultural non-integration in her case because she spoke English perfectly well. The issue of viewing English proficiency as a measure of one's Britishness is problematic, inaccurate and speaks to the larger issue of how the loyalties and belongingness of migrant communities are constantly tested. As Alexander et al. (2004) suggest in their research on the use of interpreters by ethnic minority communities in the UK,

> Speaking English, or not, has become part of a political debate that centres on issues about what it means to be a British citizen, and the ways in, and the extent to which, someone can choose to become part of a society. (Alexander et al., 2004, p. 65)

The focus on cohesion and Britishness has also had the grievous effect of shifting the emphasis even further away from a critical issue that is already underreported—that individuals who speak little or no English continue to have poor experiences of provisions in health, education and welfare. For Maria, her proficiency in English also represents her 'proficiency' in British culture, which she believes should shield her religious identity against the professional gaze when she is interacting with White practitioners who otherwise would have conflated her religiosity with her reluctance to integrate. Her proficiency in English also seems to influence her opinion about mothers who are first-generation British Pakistani immigrants, who may not speak English with sufficient proficiency and hence may not be as forthcoming when discussing their child's disability, or may simply reduce their child's disability to just a child's naughty behaviour.

> I asked her [first-generation British Pakistani mother] 'what need does your child have?' … I don't think she liked that I asked … but I think it's not up to me if the other person doesn't want to open up about themselves. Then I spoke about Aamir and his needs … but I find that in the Asian commu-

nity, mothers don't come out and say this is what their child has got, and again I think these mothers from Pakistan … from abroad … they maybe don't understand [about their child's disability] themselves. They just say they've got naughty children … they don't listen or learn and I don't think they totally understand the whole thing … it's the language … because they can't speak English, so they don't understand it … they just go ahead with what the doctors say.—Maria

Maria's remarks about first-generation immigrant mothers reflected a broader sentiment amongst mothers who were British born about those mothers who became British through transnational marriages. In the current study, British-born mothers and first-generation immigrant mothers developed their own cultural bubbles distinct from each other, which were based on their understanding of how diaspora communities should exist within British mainstream society. This entailed not only how British Pakistanis present themselves outwardly, but also more importantly how they act inwardly with regards to those aspects of their Pakistani culture that are compatible with their lived experiences and so should be retained. Not surprisingly, they each had their own experiences of exclusion which I will discuss in Chap. 6. Notably, Maria's perception of first-generation migrant mothers who lack English proficiency being less likely to integrate and less likely to understand their child's disability, is a sad reflection of the dominant narrative within mainstream literature, and perhaps also similar to Alice's perception in Chap. 1. For instance, Campbell (2015) in her USA-based research has highlighted how White professionals develop conscious and unconscious biases about the lack of education or knowledge within Black American communities, if they use a dialect other than standard American English. Unsurprisingly, Maria's perception of "*these mothers from Pakistan*" is contradictory because on the one hand it portrays mothers as being disengaged and unaware of the provisions available, whilst on the other hand they perceived as being passive partners when engaging with medical professionals.

For some mothers, forming their own cultural bubbles came at the cost of feeling that they were existing in 'no-man's land'. They were not close to the local South Asian community but they also struggled to form ties with White British families. In some instances, these cultural bubbles were also messy because on the one hand they were rooted in classism, whilst on the other hand they held an awareness that they may be seen as outsiders within their own ethnic community due to differences in their cultural and social heritage rooted in the urban/rural divide in Pakistan.

> ... you can see from the area we live in, there aren't any of our people here and my husband prefers we stay away from them [Asians]. Like those people on the other side of the city, mostly that's where they are ... My husband said you can tell the types of people who've come from Pakistan, the majority of people I've seen come from low socioeconomic class, they don't have any educational level ... living here they may have made their money, but their mentality doesn't match ours. I've met some of these people, they asked why I choose to live in the Southwest, why don't I move to the Midlands where there's a greater sense of community, but it doesn't make any difference to me.—Alina

> Sometimes I feel isolated, I've got used to that now but I moved here from a big Asian community in Northern England to a completely White area ... I did find it very difficult at first, but you adjust ... I've adjusted and I'm happy with that now, with no Asian aunties gossiping.—Kiran

This feeling of living in no-man's land was most evident in Kiran and Alina's experiences who purposely chose not to live in Pakistani neighbourhoods. They considered the mindset of the local British South Asian community to be very different from their own, but they also struggled to develop ties with their next-door neighbours and did not have any close friendships with White British families. However, the experiences of Alina and Kiran are not an anomaly; rather they are the consequence of there being a majority versus minority context within the British Pakistani diaspora itself. Nearly two-thirds of the British Pakistani population can trace their heritage to Mirpur in Azad Kashmir, and whilst the official UK Census recognises them as Pakistani, most of the younger generation identify themselves as 'ethnic Azad Kashmiris' rather than as Pakistanis (Shaw, 2001). Whilst their immigration trajectories were different, both Alina and Kiran were from Punjabi Province, which is a distinct though neighbouring territory to Azad Kashmir; moreover, their families were from major metropolitan cities in Punjab rather than from rural villages in Azad Kashmir. This divergence of background did affect the level of affinity both Alina and Kiran felt for the existing British Pakistani community where they lived, who were mostly Azad Kashmiris and were perceived by them to have different socioeconomic backgrounds, lifestyles, and occupations as well as very traditionally gendered roles for men and women. As an NHS doctor herself, Alina also felt that most of the women from the local community were not career oriented possibly because they may not have a higher educational background, and hence this was not a social network that she wanted to be associated with. Similarly, Kiran, who worked as an administrator in a mainstream nursery also expressed that she had little time or interest in forming ties with women who were different to her in terms of their worldview on gender roles and family dynamics. Whilst this flight away from British Pakistani neighbourhoods should generate greater chances of developing ties with different ethnic groups, it has also resulted in these mothers having limited informal social circles.

There were points of commonality between different cultural bubbles. For instance, when I raised the subject of non-religious Pakistani cultural traditions, all the mothers discussed this in a similar manner.

> Our culture is a headache actually. It just says, you can't do this, and you can't do that. Most of our rules aren't to do with religion, it's our culture. Especially when you live with an extended family and you live in a community dominated by Pakistanis … and you have to say to your daughter, 'you can't wear that top, it's too tight' or 'you can't wear jeans' … you're always thinking about what the neighbours are thinking and I've found myself saying to my daughter, 'what's expected from our culture is that you've got to be covered, I don't mind what you're wearing as long as you're covered' … It's about adapting to where you're living … I think it's a good factor but it's a nuisance as well because you can say to the child you can't wear that but then they'll go round the corner and then change into something that's too tight or revealing. So it's better to say, okay whatever, just wear it from home so I can see what you're wearing. So it's a nuisance.—Saira

> I don't understand why if you're living in Britain, you want to keep Pakistani culture. I mean, I like wearing shalwar kameez [traditional Pakistani garments] but I wouldn't at all want my children to wear that. If you can still practice modesty wearing jeans, then why not? And trust me, you can look shameless even in shalwar kameez, so there's no single acceptable code of dress.—Parveen

A cultural dress code is the primary feature that has classified South Asian Muslims as separatist or non-integrationist by the British mainstream media and official government discourse; ironically, this is the very feature that also presents South Asian Muslims as separatists within their own communities. Chanda-Gool (2006) suggests that cultural clothing has the potential to be a tool for social control as well as a way to assert a stronger collective identity. In her study, British Muslim and British Sikh participants discussed the unique challenges they faced in relation to their dress codes as well as the broader community expectations of them. Chanda-Gool (2006) suggests that the British South Asian diaspora face a conflicting dilemma when they arrive at their own authentic understanding of their dress code. A cultural dress code represents on the one hand a way for a community to maintain its cultural distinctiveness, whereas on the other hand to a Western audience it symbolises an inhibition of one's independence and a subjugation to primitive and oppressive values. This double gaze may often lead a diasporic community to feel torn between these two worlds.

Mothers in this study reported that they sensed a blurring of boundaries between what constituted Pakistani culture such as dressing in a traditional shalwar kameez, and the Islamic value of dressing in modest attire. Community leaders scrutinised those families where children and young people—girls in particular—wore Western clothes, silently judging them as un-Islamic even if their clothes were modest. Some mothers in my study highlighted how such cultural religiosity often created obstacles to their successful integration into British mainstream society. Notably, cultural religiosity also negatively affected the uptake by South Asian disabled families of those formal provisions and services that community members had deemed un-Islamic; this was reflected in Maria's hesitance to use respite care for her daughter because her husband was concerned that in part the community would think it was inappropriate. Unsurprisingly, the social and cultural pressure from moral policing was greater for those mothers who lived within ethnically rich neighbourhoods such as Maria, Saira and Tahira, and less for those mothers who were geographically more dispersed.

As a second-generation British-born mother, it was natural for Saira to worry that her daughter Farha would also feel torn between community and peer pressure. As I have already discussed, many mothers felt that cultural dress codes were an inconvenience. Saira's feelings capture the tension we explored earlier of framing maternal experiences through the religion versus culture paradigm. Living within a Pakistani ethnically rich neighbourhood and with her extended family and other relatives also living close by, she felt that cultural customs had become the yardstick with which the local community could police individual families. Again, mothers acted as shield for their children against such surveillance by older generations of their family members as well as community members. Saira understood that the pressure that her daughters and sons faced in terms of meeting 'cultural expectations' were not necessarily a reflection of her own values. She had made a point of letting her daughters wear whatever they were comfortable wearing, whilst also teaching her boys to do 'traditionally feminine' chores in the house; for instance, she proudly told me how all her boys knew how to cook and do laundry. She wanted to inculcate an honest and informal relationship with all her children because she recognised that they had a lot to deal with figuring out how they could exist in these contested spaces; she wanted to ensure that they would always have her to lean on if they were struggling with pressure to conform. These pressures were more consequential for Saira and her family than they were for Maham or Kiran because Saira lived in an ethnically rich neighbourhood whereas Maham and Kiran lived in a predominantly White area. Another reason why these non-religious cultural traditions that I have mentioned had more bearing on Saira and her children, could be that Saira's children were perceived as being able-bodied because they had invisible disabilities such as ADHD, depression and other mental health needs. In that sense, Saira's children still had to prove their 'cultured-ness' to community gatekeepers as they reached a marriageable age because of their perception as able-bodied young people.

It is notable that non-religious cultural traditions such as cultural dress codes were viewed as a nuisance to be imposed on children even if the mothers themselves liked to follow the very same dress code, such as wearing shalwar kameez. For instance, Parveen wore traditional Pakistani clothes at home and wore the face veil in public spaces; however, she felt that for her children modesty could be embodied by wearing Western clothes that fully covered the body. She felt that imposing a cultural dress code on her children would not only be an inconvenience, but would also serve no purpose because her children were capable of coming to their own understanding of how to be British Pakistani Muslims and whether they wanted to keep the non-religious cultural traditions as part of their own cultural bubbles.

CULTURAL COMPETENCY IN COMMUNITY AND INSTITUTIONAL SPACES

In distinguishing their own cultural bubbles from the overarching image and cultural values of the British Pakistani community, the mothers in the current study began to highlight the disadvantages of such community cultural values and the justification for forming their own cultural bubbles within their communities. For instance, the mothers discussed the lack of value placed on inclusion within wider British Pakistani community. It is interesting to note that Pakistani culture does embody strong collectivist and familial values; however, it is possible that due to the lack of understanding and awareness around disability that disabled families are excluded from the community's collectivist support structures. As already discussed in the Chap. 3, this lack of an inclusive ethos for disabled families appeared to be most visible during religious or cultural events where even the most basic accommodations were missing, or where the need to provide the necessary accommodations was met with indifference or annoyance. The lack of an inclusive ethos was also visible in attitudes of community members.

> ... they [South Asian community] don't listen ... they don't talk in a good manner and view us negatively ... in the beginning when I was taking Farrukh out, some Asian people were making fun, laughing at him in front of me ... I trust British society more than I trust our own community ... Sometimes, they don't approve of us.—Tahira

> We're very rigid people, we don't have flexibility towards mothers [who have children] with disability, we don't have the sense to do some social work, that's never been taught to us. So expecting anything from the Pakistani or Asian community isn't something I'd suggest because you won't get it. You can't expect an overnight change but they're not even that concerned about it to be honest.—Alina

There was strong agreement amongst all the mothers in the current study that at a community level, an understanding of what constitutes inclusion, and specifically inclusion of children with disability, was missing. The mothers commented that even where community attitudes were not exclusionary, their views were based on notions of giving charity. Even with little to no informal support, the mothers chose not to include such community members in their cultural bubbles because they did not want to be perceived as being needy and incompetent in supporting their disabled child. For instance, Tahira also lived in a predominantly British Pakistani neighbourhood and had her extended family living nearby, and yet she did not utilise them for any informal care. This was based on her previous experiences where she had realised that she and her disabled child were not welcome by her extended family, and were viewed more as an inconvenience. This led Tahira to sacrifice a lot of her own personal time and energy to ensure that her reliance on the informal support network was minimal.

For Alina, this was not simply a case of deliberate exclusion by the British Pakistani community but rather the result of not being taught the values of inclusion, social work and rights of disabled children early on. A few scholars have posited that this is down to the imposition of Western interpretations of inclusion or social work onto pockets of Global South communities, who may not relate to or feel that these versions represent their cultural values because they have their own collectivist and familial values that embody inclusion and social work. For instance, Miles (1996) suggests that the Eurocentric concept of 'disability' and social models of disability differ greatly from how South Asian culture has traditionally conceptualised and understood disability, which has resulted in disability and disabled families often being stigmatised by the South Asian community. Whilst Miles primarily focuses on disability planning in South Asian countries rather than within South Asian diasporic communities in the UK, there is little attention paid to how caregiving and familial ties that are commonly associated with South Asian collectivist values are currently being utilised to promote inclusion. Alina's comment was quite revealing because it suggests that the British Pakistani community does not really 'give' back, and that as a community it did not place any value on caregiving as a formal or informal service. Yet within the city, there were a few major British Pakistani and British Bangladeshi support centres for disabled children, young people and adults. One of these centres, Anokha, was also the community centre that had acted as a gatekeeper, giving me initial access to some of the mothers who had gone onto become my research participants. It is possible that Alina's experiences may also have been mediated by her immigrant status; she was still in the process of becoming a UK citizen and so had no recourse to public funds, having thus far had to pay for all the provisions that Imran was receiving. In having to arrange and pay for everything herself, she saw little benefit in engaging with the community centre.

With each conversation, it became increasingly clear that Alina and some of the other mothers in the current study did not associate with Anokha, nor did they find them helpful despite the services and support they offered. The Anokha Community Centre prided itself on hosting all major Pakistani cultural and Muslim religious events, always in a disabled accessible manner, and often also organised cross-cultural events such as

Christmas and Easter as well as other non-Islamic festivals such as Diwali, Holi and Vaisakhi. Some mothers revealed that whilst this community centre was central to providing information on respite and other services, its organisers had become 'self-declared community leaders' (Sanghera & Thapar-Bjorkert, 2008) who wielded considerable influence and power over community members. During my data collection phase and my inter-action with this community centre, it was evident that the centre organisers held implicit biases against certain families in the community and viewed some families more favourably. In becoming self-declared community leaders, Anokha made a point of knowing everybody in the community, but a negative and harmful by-product of this community engagement entailed knowledge of the intimate details of the private lives of community members. Indeed, some mothers like Kiran needed firm assurances from me that I did not represent the community support centre when I first approached them for interviews. Kiran was worried about her intimate details being shared with the community centre, which stemmed from her previous experiences where she felt that Anokha had shared her personal and family problems within mutual social circles.

> It's based on experience ... with them [Anokha] confidentiality is very low, we just don't know what confidentiality is in the Asian community ... I've learnt that ...—Kiran

This mistrust amongst some mothers affected my credibility as a researcher, despite the fact that I was not a 'friend' or 'associate' of Anokha (Crowhurst, 2013). It was only after repeated assurances that her personal accounts and narratives would not be shared with Anokha organisers, did Kiran consent to participate in the current study. I also observed how the organisers at Anokha discouraged me from approaching some families, advising me that they were not 'right' for my research. This was a major reason why I decided to employ network sampling to make initial contact with those mothers who might not be viewed favourably by the organisers at Anokha.

It is difficult to quantify whether membership of British Pakistani communal spaces like the Anokha Community Centre offers any tangible or intangible benefits to these mothers. Some mothers remained keen to attend religious events held by the community centre such as Eid, and cultural events such as Pakistani bridal festivals. However, with time it became evident that for many mothers, the culture within this community centre had come to represent the less favourable side of broader British Pakistani culture, one where familial and collectivist values had become conflated with overt gossiping, a lack of respect for confidentiality and the policing of individual family practices who did not conform to the community organisers' view of how British Pakistani families should behave. In contrast to this, the mothers also observed the institutional culture of formal services in the UK whilst attending health clinics or formal support sessions organised by key workers, a culture that to them exuded inclusivity and tolerance which in turn came to represent the culture of mainstream British society. What is interesting is that mothers associated formal institutional spaces with White people, whilst informal communal spaces became associated with *desi* British Pakistani people. Irrespective of whether mothers were British-born or first-generation immigrants, they all valued privacy, confidentiality, professionalism, tolerance and empathy as being critical to their experiences of inclusion; notably, mothers perceived that these values were most reflected within formal institutionalised White spaces. Some mothers expressed hesitance in approaching the community centre lest their personal matters were leaked into the wider community orbit and everyone in the community becoming aware of their difficulties. Perhaps this maternal hesitance to approach these informal communal spaces was magnified because they had formed their own small cultural bubbles that had their own set of values and practices, and which might have been disapproved of by Anokha's self-declared community leaders. For instance, Alina commented that she felt uncomfortable attending such spaces because she felt judged by Anokha's conservative dress code, which frowned upon her 'sleeveless kameez'; she also believed the sessions that Anokha offered did not represent her family practices or values.

This association of professionalism and confidentiality with White people is a telling finding in that, on the surface at least, it appears to go against what the majority of extant literature suggests—the need for White professionals to undertake cultural competency training when engaging with minoritised communities. However, it is important to dissect further what these maternal experiences actually inform us here. All formal institutions in this country such as schools, hospitals and social care services are bound by a code of practice that mandates professionalism, confidentiality and anonymity in compliance with General Data Protection Regulation (GDPR), and the tolerance and inclusion of diverse views in compliance with the Equality Act (2010). These codes of conduct are designed to help these formal institutions to become more responsible and accountable towards their end users. It is not surprising that Parveen felt that children's rights were an integral part of British mainstream society,

> If someone over here abuses or mistreats a child, then it's taken seriously ...
> it will be rare case [where] such an injustice has been overlooked or the
> offender escapes the system.—Parveen

Parveen is referring to the presence of safeguarding and other macro laws in the UK which prevent discrimination and harm all children and young people, rather than the general attitudes of White British society. It is not only inaccurate but also unhelpful to conflate British society as being generally more accepting and appropriate, despite evidence to the contrary which points to rising xenophobic and exclusionary attitudes against minorities within Britain potentially having worsened in recent years. Indeed, according to one recent survey by YouGov (2018), UK adults hold very negative attitudes about British Pakistanis and British Bangladeshis. However, Parveen's individual lived experiences suggested that for her, existing communal spaces were more exclusionary, and it is difficult to say that they were not. The impact of her exclusion from this informal *desi* communal space was far more personal and real than her exclusion from any institutional formal structures that nonetheless offered her superior support, even if they may have viewed her visible 'Muslimness' as at odds with the British way of life.

If we examine informal community spaces more closely, they operate very differently to formal institutional spaces. Many community spaces are wholly reliant on individual volunteers in order to operate, who can give their time and resources into providing a facilitative environment for others. In some instances, self-declared community leaders within British Muslim communities such as Muslim imams in local mosques, and first-generation community elders, both men and women can take over this role. Whilst communal spaces abide by macro laws such as legal requirements that ensure these spaces adhere to general safeguarding laws, they often lack the funding to provide specialist training to their volunteers or to establish structures within the organisation that ensure everyone is included. These spaces, however, should not be dismissed as being exclusionary and of little value to disabled families; after all, Anokha, like other local community centres still offer support to South Asian disabled families, even if this support is selective. Rather, as bell hooks (2000) suggests, these informal spaces present the potential to transform the experiences of support for disabled families by challenging patriarchal structures that harm the very families they seek to serve.

This is not an easy feat to accomplish because it requires existing community leaders to unlearn the internalising narrative of how immigrants should assimilate, and to establish new codes of conduct that are driven by the needs and concerns of diverse families from the community. This change has to reflect intergenerational concerns and the views of newly arrived families from Pakistan such as Alina's, who have very different demographic backgrounds. In some sense, communal spaces or centres need to develop and establish their own cultural competency frameworks that acknowledge how multiple forms of oppression can disadvantage different families. For instance, during our conversations, Tahira revealed her concerns about approaching an informal communal space such as Anokha for support and respite, or to share her personal experiences such as how her labour was divided between caring for her disabled husband, her disabled child and her ageing parent in laws. She told me that she needed to be assured beforehand that her very personal problems would not be shared with the wider community, or that she would not be merely advised that her problems were *ghar ka mamla* (a personal matter)[3] which she must learn to deal with it herself. Therefore, informal communal spaces must be able to separate and expel the harmful patriarchal elements of Pakistani culture from the positive familial values such as collective care, ethics of caregiving, etiquette and respect for maternal labour. This is not implausible considering the fact that these community spaces were initially formed to compensate for the gaps and failures of formal institutions in engaging with different minority communities.

It must also be noted that formal institutionalised spaces have invested heavily in cultural competency training in recent decades in order to reduce these service gaps and failures. Campbell (2015) adds that although these cultural competency models have developed some level of sensitivity to help facilitate professionals who engage with minority families, nonetheless, many White professionals still fail to recognise minority families as racialised beings who experience multiple forms of oppression. This is likely explained, she suggests, by the fact that these cultural competency

---

[3] *Ghar ka mamla* is a patriarchal response to different ways violence is enacted within families. It means this is a private matter and does not merit a public response.

frameworks build up professional understanding of cultural awareness and knowledge based upon *"academic fields such as anthropology, or sociology, which continue to encourage stereotypes and assumptions of cultural, racial and ethnic groups"* (p. 14). For instance, in 2018 the Citizens Advice Bureau (CAB),[4] an independent organisation that provides free, impartial and confidential advice to the public on matters relating to legal, consumer and money problems were criticised for including extremely stereotypical and racist information about ethnic minoritised communities as part of their continuing professional development sessions on how to interact with minority communities in the UK. The supposed 'common traits' that the CAB (Mohdin, 2019) sessions listed about British ethnic minority communities included having large families, a distrust of British authorities, early marriages, a focus on honour and shame, and keeping cash at home among many other examples. It is not difficult to imagine probably very similar negative and problematic assumptions and stereotypes about South Asians, that persuaded teachers to provide brochures warning of the dangers of forced marriage to Saira when she suggested that she and her family were considering visiting Pakistan during the summer holiday, without realising how damaging this was for their relationship with Saira. It is also entirely likely that stereotypical assumptions such as these that led to Tahira being distrusted by social services and investigated for a potential suicide attempt, rather than being offered the support that she had requested.

The fact that mothers in the current study equated institutional spaces with White people being more professional and trustworthy, reveals how internalised the colonialist mindset can be within minoritised communities in the UK. Additionally, it also demonstrates how some minorities might conform to colonialist constructions or reductive discourses if they can see some social and economic gains for their children (Brah, 1996). On the one hand, they are aware that they are perceived as a problem by mainstream British society, yet on the other hand, they know that they have to perform to dominant scripts in order to be considered deserving of the services and provisions that they need. Mothers like Saira felt that they were continuously having to prove that they were not akin to their stereotypes, and that at some point White professionals also need to step up and recognise the inter-categorical positionings that minority disabled

---

[4] https://www.theguardian.com/uk-news/2019/aug/14/citizens-advice-training-document-propagated-racist-stereotypes

families occupy. The reductive analysis of problems that minority disabled children and families face have often resulted in the poor or ineffective provisions and interventions directed at them.

> I would expect teachers to be better trained for the awareness around young children, teenagers ... more about ethnic minorities, about our culture, about our background, about our history, about the fact that this child is growing up in a mixed community and is being pulled in every direction.—Saira

The consequences of being problematized by mainstream institutional spaces as well as of being culturally policed by communal spaces, often meant that mothers were left with no support system. In some instances, mothers have been fortunate enough to start up their own communal spaces that support mothers like them. For instance, Parveen had over the years created her own space at her home where other mums would meet up weekly, and she would offer support to them in terms of self-care, as well as create a better understanding of how gendered caregiving roles in the home could be detrimental to maternal health and wellbeing. This creation of self-help communal spaces by mothers themselves has also redefined certain aspects of culture that mothers of disabled children want to keep and those aspects they want to discard, and in that process develop their authentic understanding of their cultural selves. It also reflects gender costs for mothers where they have to actively create a space for themselves because an existing space is exclusionary (McLoughlin et al., 2014).

REFERENCES

Akbar, S., & Woods, K. (2020). Understanding Pakistani parents' experience of having a child with special educational needs and disability (SEND) in England. *European Journal of Special Needs Education, 35*(5), 663–678.

Alexander, C., Edwards, R., Temple, B., Kanani, U., Zhuang, L., Miah, M., & Sam, A. (2004). *Access to services with interpreters: User views*. Joseph Rowntree Foundation.

Azmi, S., Hatton, C., Emerson, E., & Caine, A. (1997). Listening to adolescents and adults with disabilities from south Asian communities. *Journal of Applied Research in Intellectual Disabilities, 10*, 250–263.

bell hooks. (2000). *Feminist theory: From margin to center*. Pluto Press.

Bhardwaj, A. K., Forrester-Jones, R. V., & Murphy, G. H. (2018). Social networks of adults with an intellectual disability from south Asian and white communities in the United Kingdom: A comparison. *Journal of Applied Research in Intellectual Disabilities, 31*(2), 253–264.

Bhatti, G. (1999). *Asian children at home and at school: An ethnographic study.* Routledge.

Bolognani, M. (2009). 'These girls want to get married as well'. Normality, double deviance and reintegration amongst British Pakistani women. In V. S. Kalra (Ed.), *Pakistani diasporas: Culture, conflict and change* (pp. 150–166). Oxford University Press.

Bolognani, M., & Mellor, J. (2012). British Pakistani women's use of the 'religion versus culture' contrast: A critical analysis. *Culture and Religion, 13*(2), 211–226.

Brah, A. (1996). *Cartographies of diaspora: Contesting identities.* Routledge.

Campbell, E. L. (2015). Transitioning from a model of cultural competency toward an inclusive pedagogy of "racial competency" using critical race theory. *Journal of Social Welfare and Human Rights, 3*(1), 1–16.

Chanda-Gool, S. (2006). *South Asian communities: Catalysts for educational change.* Trentham Books.

Crowhurst, I. (2013). The fallacy of the instrumental gate? Contextualising the process of gaining access through gatekeepers. *International Journal of Social Research Methodology, 16*(6), 463–475.

Crozier, G., & Davies, J. (2008). 'The trouble is they don't mix': Self-segregation or enforced exclusion? *Race Ethnicity and Education, 11*(3), 285–301.

Equality Act. (2010). *London, UK: The stationery office limited under the authority and superintendent of Carol Tullo, controller of her majesty's stationery office and queen's printer of acts of parliament.* Retrieved December 15, 2020, from http://www.legislation.gov.uk/ukpga/2010/15/pdfs/ukpga_20100015_en.pdf

Harry, B. (2002). Trends and issues in serving culturally diverse families of children with disabilities. *The Journal of Special Education, 36*(3), 132–140.

Hatton, C., Akram, Y., Shah, R., Robertson, J., & Emerson, E. (2004). *Supporting south Asian families with a child with severe disabilities.* Jessica Kingsley Publishers.

Mamdani, M. (2005). *Good Muslim, bad Muslim: America, the cold war and the roots of terror.* First Three Leaves Press Edition.

McLoughlin, S., Gould, W., Kabir, A. J., & Tomalin, E. (Eds.). (2014). *Writing the City in British Asian diasporas.* Routledge.

Miles, M. (1996). Community, individual or information development? Dilemmas of concept and culture in south Asian disability planning. *Disability & society, 11*(4), 485–500.

Mohdin, A. (2019, August 14). Citizens Advice training document 'propagated racist stereotypes'. *The Guardian.* Retrieved December 15, 2020, from https://www.theguardian.com/uk-news/2019/aug/14/citizens-advice-training-document-propagated-racist-stereotypes

Rizvi, S. (2015). Exploring British Pakistani mothers' perception of their child with disability: Insights from a UK context. *Journal of Research in Special Educational Needs.* https://doi.org/10.1111/1471-3802.12111

Sanghera, G. S., & Thapar-Bjorkert, S. (2008). Methodological dilemmas: Gatekeepers and positionality in Bradford. *Ethnic and Racial Studies, 31*(3), 543–562.

Shaw, A. (2001). Kinship, cultural preference and immigration: Consanguineous marriage among British Pakistanis. *Journal of the Royal Anthropological Institute, 7*(2), 315–334.

YouGov. (2018). *Where the public stands on immigration.* Retrieved December 15, 2020, from https://yougov.co.uk/topics/politics/articles-reports/2018/04/27/where-public-stands-immigration

CHAPTER 6

Englistan and Citizenship

The tensions between immigrant communities and their host society have often focused on feelings of being torn between two countries, of split allegiances, of having to continuously feel temporality in belongingness and of imagining a return to their country of origin (Sayad, 2004; Bhatti, 1999). The mothers in this research have tried to break away from this static and binary representation and decided that whilst they may not be accepted as fully British, nonetheless, they have made Englistan their permanent abode.

> Being condemned to refer simultaneously to two societies, emigrants dream without noticing the contradiction of combining the incompatible advantages of two conflicting choices. At times, they idealise France, and would like it to have, in addition to the advantages it gives them (a stable job, a wage, etc.), that other quality of being a second land of their birth- which would be enough to transfigure the relationship and to magically transform all the reasons for dissatisfaction they experience in France. At other times they idealise Algeria in their dreams or after spending time there during their annual holidays. They want it to correspond to an idealised France ... (Sayad, 2004, p. 58)

In the above quote, Sayad (2004) writes about the predicament of French Algerians, who are always willing to imagine France as the idealised home that it should be, which embraces their 'Algerianness' as well as

S. Rizvi, *Undoing Whiteness in Disability Studies*, https://doi.org/10.1007/978-3-030-79573-3_6

retaining those aspects of France that they find so desirous in a homeland. However, Sayad posits that French Algerians are stuck in no man's land because they have been disenfranchised and minoritised by successive French governments, their right to exist in France always viewed from an economic 'cost versus benefit' analysis by the French government, the media and mainstream society. When we examine literature on British South Asians and their experiences of belongingness to Britain, we come across notions of the 'myth of return'—the idea that once minority communities have saved enough money, and secured a good future for their children, they will migrate back to their country of origin. Dahya (1974) posits that:

> The immigrants come with the firm intention of returning home where they hope to enjoy the fruits of their labour in retirement ... (They) consider themselves to be transients and not settlers. However, this is not to imply that the immigrants, or any significant number of them, will in fact return home. (Dahya, 1974, p. 83)

Dahya (1974) suggests that the myth of return does not entail an actual physical return of immigrants to their country of origin, rather it represents a hope or thought of living out their retirement in relative financial security in their 'homeland'. However, the myth of return also perpetuates a harmful narrative that minority communities have not 'done' enough to prove their worth as loyal citizens, who may have 'cheated the system' in the UK so that they can transfer the benefits to their country of origin (Sayad, 2004). Whilst Dahya (1974) utilises a romanticised lens to explore the myth of return, I prescribe to the starkly realist lens in order to view this transient belongingness; it is not the consequence of a failure of immigrants to integrate into the UK but rather because they are never allowed to feel 'at home' in the first place.

This exclusion is imposed by a country's immigration policy, its political and media discourse, and the way it engages in a cost versus benefit analysis of its ethnic minority communities. This othering discourse paints all ethnic minority communities as immigrants, as outsiders regardless of their immigration trajectories and their generational ties to the UK and the British Empire, and regardless of their efforts to prove their loyalty to the UK. It is a vicious discourse because it seeks to erase the very ground on which communities build their lives in the UK, unless they perform to certain nationalist scripts. The Windrush Scandal is one tragic example;

the 'Windrush generation'[1] refers to families who arrived in the UK from Caribbean countries between 1948 and 1971 (Williams, 2020). This immigration policy was instituted to fill labour shortages after the World War II as the UK economy boomed, and mainly invited members of Empire nations to immigrate to the UK to work in and build up the health and transport sectors. The British Government placed adverts that declared, "*Come to the Mother Country! The Mother Country Needs You!*" (Williams, 2020) that encouraged many families in the Caribbean to consider immigration to the UK as a good opportunity for their families. Families from Empire countries were considered natural citizens since they were British subjects, and were by law allowed to live permanently in the UK; notably, they were not required to submit any paperwork for their immigration applications. In 2017, it emerged that successive Conservative Governments had been using its "*hostile environment*" immigration policy to illegally detain, deport and deny legal rights to those Caribbean families who could not show proof of citizenship (Grierson, 2018). Since these British Caribbean families had not been required to have paperwork when they first arrived, they were trapped by a government policy that was wrongfully detaining and deporting them.

The change in immigration policy over the last few decades has made it extremely difficult for families outside of the EU to immigrate to the UK. In addition, successive British governments have increased their scrutiny of ethnic minority communities, particularly their immigration status even if they have resided in Britain for generations. Immigration policies have also increasingly shifted from an economic discourse to a political one, endorsed by the current rise in populism across Britain, the US and many European countries. The recent case of the British Bangladeshi woman, Shamima Begum, who in 2015 as a 15-year-old travelled to Syria after being groomed online by ISIS terrorist Yago Riedijk,[2] highlights how governments can use their power to render minorities—in this case, a legal minor who was at risk of online radicalisation—stateless, revoking their citizenship without any consideration of their rights. The Windrush Scandal and Shamima Begum's case present a changing of immigration

---

[1] The Windrush generation is named after the ship *HMT Empire Windrush* which brought one of the first significant groups of Caribbean people to the UK, mooring at Tilbury Docks on June 21 1948.

[2] After arriving in Syria, Shamima married Riedijk and gave birth to three children, all of whom are now dead. She was later discovered in a Syrian refugee camp in 2019 by a British journalist and was denied re-entry to the UK. In 2020, the Court of Appeal ruled that Shamima could come back to the UK to challenge her UK citizenship revocation in court.

policy implementation that instils fear and uncertainty about the rights of ethnic minorities to live in the UK.

The debate about who counts as a citizen is multi-layered, being deeply gendered, ableist and racialised. Meekosha and Dowse (1997) posit that the renewed focus on citizenship by liberal democracies seeks to build a nationalistic community through a rights/duties discourse, imposing "*notions of active and responsible citizens who are less dependent on state and less able therefore to make claims against the common weal*" (p. 53). This agenda actively excludes those groups which threaten its cohesiveness, such as groups at intersections of immigration status (e.g. refugees and asylum seekers), gender (i.e. women), ability (i.e. disabled), and racial and ethnic identity (e.g. Muslims or other people of colour). They are perceived as neither "*productive nor potential citizens*" (Meekosha & Dowse, 1997, p. 59), and are portrayed as 'threats' to liberal democracies. As previously discussed in the Chaps. 3 and 4, disability and consanguinity within the British South Asian community has been portrayed as an issue that needs to be 'contained' because these communities are becoming a burden on existing health, education and social care services. As posited earlier, the rights of British White disabled communities are perceived as being separate from the rights of ethnic minority disabled communities, which are 'hyper-visibilised' and discussed from the standpoint of the national economic agenda, and in doing so, are subtly re-writing who counts as a citizen. Is it possible to express grievances against current state of provisions and interventions for minority disabled families without being perceived as 'welfare queens or tourists' or as a problem to be taken care of? They are also 'invisibilised' because provisions and interventions aimed at improving education, health and social care are not designed with these communities in mind. In this chapter, therefore, I unpack what mothers in the current study refer to as their homeland and where they feel comfortable and secure enough as citizens to ask for what is rightfully theirs by law.

SHIKWA[3] WITH PAKISTAN

As mentioned in the Chap. 2, mothers in this research possessed divergent immigration trajectories. Some were first-generation migrants through transnational marriages (i.e. Tahira, Maham and Parveen), whereas others

---

[3] Shikwa is an Urdu word that translates to complaints. But it is not to be read as complaints in a formal relationship, but as lamenting with a loved one.

had migrated to the UK to ensure a better life for their disabled child (i.e. Alina), and some were British-born second-generation immigrants (i.e. Kiran, Maria, Shehnaz and Saira). Even with their varying experiences, all mothers nonetheless referred to Pakistan in the context of being a 'third world country' which could never be a permanent abode for their disabled children. Maternal views were particularly strong regarding Pakistan's lack of appropriate provisions, mainstream school places, teacher training and specialist support for disabled children.

First-generation migrant mothers appeared to be fighting for their child's inclusion at an intrinsic level, rather than just inclusion within mainstream schools. They constantly rationalised their negative experiences of settling as immigrants with the overall benefit that the UK had provided their child, which reflected their decision to make Britain their permanent home. For my participants, there was no "*myth of return*" or notion of going "*back home*" unlike parents in Bhatti's (1999, p. 9) study. First-generation migrant mothers in my research did not even revisit Pakistan on holiday because of the perception—and in some cases the reality—of a lack of appropriate provisions and infrastructure. Nor did they want to give their disabled children the experience of forging stronger connections with their kin in Pakistan in case their relatives did not have an accurate understanding of disability, or due to the inconvenience of hearing unsolicited advice about child-rearing.

The study by Goodley et al. (2013)[4] with three British Pakistani families reveals how they viewed Pakistan as an "*imagined community*" (p. 6). Outwardly, they celebrated certain aspects of culture such as watching Pakistani TV drama serials or wearing traditional Pakistani shalwar kameez dress; however, they were also deeply affected by their experiences of supporting their disabled child. Mothers in my research did not romanticise this 'imagined community', and were not as keen to adhere to their Pakistani cultural identities as the families in the study by Goodley et al. (2013) did. This also reflects a keenness on the part of mothers generally in this current research to establish meaningful and long-lasting relationships with educational and other services providers in Britain. Having had first-hand experience of growing up in Pakistan, some first-generation

---

[4] Study by Goodley et al. (2013) employed a range of qualitative methods to look at the current experiences of disabled children and young people post-Blair. As part of their study, they also interviewed three British Pakistani mothers with disabled children.

immigrant mothers highlighted several obstacles that their child could have experienced if they had been living in Pakistan.

> There's no specialist provision, people don't know about autism, there are no specialist consultants or behaviour analysts or speech and language therapists. There are no specially trained people … they just wanted him to be in a school for handicapped children, or deaf and dumb or physically disabled children but I couldn't think of sending my child to that kind of place.—Alina

Alina had migrated to the UK solely to ensure better support for her son, Imran. As discussed in earlier, Alina researched about her eligibility to work in different countries and found that whilst Canada and the US had better provisions, the UK would take the least amount of time in acknowledging and validating her medical degree and experience. In Canada and USA, she would have had to sit for medical licensing examinations to be eligible for work as a doctor. Her decision to choose Britain was not straightforward, despite her husband already working in this country. It meant leaving behind a life of higher social status in Pakistan as well as family support, so that Imran could have a fair chance of being supported in school settings.

> You don't get these things [SEN provisions] in the third world, in India or Pakistan. This is everything for me, it's what forces me to stay here, otherwise if I look at myself, I've become a servant. In Pakistan, you have servants, drivers, you don't need to shop for your own groceries … your parents come live with you and help, your in-laws can also help, but here you're on your own. You have to do everything yourself … After so many years living in Pakistan, I never thought I'd be living abroad. I was happy there with my career, and post-graduation I'd thought I'd leave my job and just do a few hours of private practice. Here, I've started from a junior level again … below the level I was working at over there.—Alina

Alina's experiences are important to challenging the static and problematic representation of immigrants as benefit scroungers and welfare tourists, as she reflected on her past lifestyle in Pakistan, where she had servants, a driver, family support as well as her stable career trajectory as a doctor. Back in Islamabad, she had opened up her own private practice which allowed flexible working hours to support her mothering responsibilities. Her home in Islamabad was large enough for her parents to move in with her so that they could support her. There was a sense of lamentation when Alina reminisced about her lived experiences in Pakistan; however, the promise of material support for Imran once her British citizenship was confirmed had assured her that her decision to come to this country was correct and worthwhile. Nonetheless, she missed her parents and running her own medical practice. She had had to accept work in a junior position in the NHS and start working her way up. Despite categorising Pakistan as a *"third world"* country, she felt like a first-class citizen in Pakistan, whereas in Britain she felt like a third-class citizen due to her temporary status. She stated that, *"This is everything for me, it's what forces me to stay here, otherwise if I look at myself, I've become a servant"*. The *"everything"* that Alina refers to is the possibility that her son, Imran can exercise his full UK citizenship rights in the future in the form of access to provisions, even if it comes at the cost of a temporary loss of her own citizenship and sense of belongingness. By coming to the UK, Alina recognised that Imran could do simple activities such as going to school, shops, cinemas and the park because of an existing accessibility infrastructure, something that only able-bodied people could access in Pakistan. There was also a sense of disappointment in her reflections, knowing that whilst Pakistan has its charms it did not have a welfare system or inclusive policies like Britain that would enable Imran to be supported within mainstream settings.

Alina's perception of Pakistan as a *"third world"* country was not an exception, rather all the mothers in my research, whether or not they had visited Pakistan, had a perception that Pakistan lacked the basic infrastructure needed to support disabled children and young people. This was an interesting reflection because all the mothers discussed how there is greater freedom to do anything they want in Pakistan, a chance to enjoy a better lifestyle in terms of eating out, and cultural and religious networks; nonetheless, they still cited the lack of inclusive educational placements and provisions as reasons for never considering living in Pakistan. For Alina specifically, it was also an issue of inappropriate placement; living in Islamabad, the capital of Pakistan, contained a few specialist schools for deaf and physically disabled children but Alina felt that Imran deserved a chance within mainstream settings.

> These kids [disabled children] never come into mainstream schools ... either they are kept at home or they're straightaway sent to a special placement ... they wanted Imran to be in a school for handicapped children who were

deaf or physically disabled, but I couldn't think of sending my child to that kind of place … also there is no planning about the future … the concept we have in our minds about special needs children [in Pakistan] is either one of the parents are stuck with them or they'll have a dedicated servant or maid for them if they can afford it.—Alina

Interestingly, despite Alina's comment above, Imran was still attending a special school in the UK. In fact, whilst many mothers commented on absence of inclusion within mainstream schools in Pakistan, their child was in a special school placement in the UK. There are many reasons for this perception that mothers like Alina displayed. Firstly, specialist schools are very few in number in Pakistan, located mainly in the larger urban centres and are chronically underequipped. In a recent report by Hafeez (2020), he found that there are less than 200 special education institutes serving the whole of Pakistan, a country of over 207 million[5] people, and that only 5 per cent of disabled children are in some form of special school placement and the remaining disabled children are out of education altogether. Low levels of government interest and the resulting lack of budgetary allocation means that there is poor infrastructure, undertrained teachers, little or no accessibility within small towns and villages, and inconsistency of policy across different provinces. This is unfortunately characteristic of many countries in the Global South that have been affected by crippling foreign debts, economic sanctions, military interventions, and unstable central governments (Meekosha & Soldatic, 2011). Like Pakistan, many countries in the Global South also rely on international non-governmental charities for their special education provisions. Whilst this foreign aid and support is critical to compensate for the lack of local government support, it creates tensions around whether health or education is given priority as well as around the long-term sustainability of international projects (Miles & Singal, 2010). This has implications for disabled families in Pakistan in that there are fewer affordable services available to families from lower socioeconomic backgrounds, with a greater reliance on the private and charity sectors. Moreover, special schools operated by non-governmental organisations (NGOs) and private companies often lack a specialist focus, leading to children with different types of disability being taught in the same way.

---

[5] Population Census of Pakistan, http://www.pbs.gov.pk/content/population-census

A report by UNICEF (2003) on inclusive education in Pakistan suggested that the pedagogies employed in mainstream Pakistani schools were based on rote learning and copying, which confirms the concerns of mothers in my study. Moreover, there were significant differences in the methods and quality of teaching between mainstream and special schools in Pakistan, and an overall lack of inclusive schools in general. Poor discipline was not tolerated and corporal punishment was still widely practiced, potentially worrying for parents of children with SEMH needs; moreover, children with visible types of disability were not even enrolled by mainstream schools (UNICEF, 2003). Special schools in Pakistan were found to lack the expertise and resources needed to address children's individual needs and to support learning. Therefore, whilst non-disabled families could return to Pakistan in the full knowledge that their children would not be unduly disadvantaged educationally by attending mainstream schools, unfortunately, this was not an option for mothers in my study. The implications of the UNICEF (2003) report for my research participants was that they knew their child would not be supported within special schools in Pakistan, and would be refused admission to mainstream schools as was the case with Alina.

Mothers in my study also recognised that Pakistan does not have a welfare system, and that the government only provided very basic services which were entirely unsuitable for their children's needs. This meant that families would have to pay privately for specialist medical, social and educational expenses for their child, which would be beyond the socioeconomic means of some mothers in my sample. This was evident in Tahira's experiences,

> I'm not going back to Pakistan because of my son … I always think that if I was in Pakistan, I wouldn't have these facilities. My son might not get these kind of facilities and he might not get taken care of by teachers or the government.—Tahira

> It's quite an upsetting fact, but this is the reality. We used to live in a remote rural village [in Pakistan], so even if there were special schools available, they'd be in major cities or urban areas.—Parveen

Mothers who were born and educated in Pakistan and had migrated to the UK were acutely aware that the Pakistani education system lacked appropriate infrastructure to support disabled children. There was no 'going back', a decision that was based solely on the quality of educational and other provisions available in the UK. There was also a sense of a willingness to let go of their previous roots and their family ties weaved into their narratives, if it meant that their disabled children could have a better life in Britain.

> My dad lives in Pakistan and he's in a very serious condition right now, on his deathbed ... the doctors have given up on him. It's been seven years since I last visited Pakistan and I need to pay my last respects to him, but because of Farrukh I don't want to go. I've said that if Farrukh is allowed to travel with me even for a week, then I'll take him with me ... but the doctors have said not to take him there. They said he's poorly here [Britain], so he'll be worse off in Pakistan ... My heart wishes to go to Pakistan but then I stop myself because I don't want to leave Farrukh alone.—Tahira

> I took Sehr once to Pakistan, actually I was unable to visit Pakistan for 10 years because of her and she had an i-card [disability identification card]. I became very isolated, you can't live a normal life with a disabled child because your whole family life changes ... when I went back home [Pakistan], my friend told me 'it feels like you've disappeared from our lives'.—Parveen

For Tahira, Alina and Parveen, it meant missing out on and not being able to perform familial duties back in Pakistan. Alina relayed to me that her father's health had declined in recent years and he had been diagnosed with lymphoma; she wanted to bring him over to Britain but since her own legal status was temporary, she could not act as his visa sponsor. For Tahira, as a result of Farrukh's PEG and other complications, she was unwilling to travel alone to Pakistan and so could not visit her dying father to pay her final respects; doctors were hesitant to give her approval to take Farrukh with her to Pakistan, as they were unsure about the level of support there. All first-generation migrant mothers in the current research suggested that they had been unable to revisit Pakistan in years, and that travelling back with a disabled child—even if it was for a short trip—would be immensely challenging. Their reluctance to travel back to Pakistan was also a consequence of needing to meet the residential requirements of becoming British citizens, in case they were forced to restart their immigration process from the beginning. This ultimately meant that the familial support systems available to British-born mothers of disabled children, by virtue of having their parents and siblings living in the same country, was entirely lacking for first-generation migrant mothers.

Rather, first-generation migrant mothers came to the UK and settled immediately into their husband's joint-family systems who often expected traditional gendered roles from their immigrant *bahus* (daughters-in-law), or alternatively had to set up their home without any access to any formal or informal support systems. All of this suggests that first-generation migrant mothers of disabled children underwent a tumultuous process to become British citizens. After having made such a life-changing decision to migrate to the UK, I found them more likely to rationalise their decisions in order to ultimately be content. At one point in time, they all occupied a temporary legal status in the UK which limited their rights to access formal and informal support systems, all in the hope that their child could exercise their British citizenship in the future.

All mothers demonstrated more faith in the expertise and training of British teachers and other professionals who worked with disabled children, as compared to teachers in Pakistan. Notably, in spite of Britain's welfare support, mothers were also deeply aware of the financial cost for raising a disabled child in the UK, and so were concerned about having to pay these costs themselves in Pakistan. Hussain et al. (2002) also found that their participants stated that support was lacking in Pakistan and India, however, parents in their study in particular still valued trips to their homeland.

Another reason why mothers in my study valued the UK and had settled here was that they thought disability was highly stigmatised in Pakistan and viewed through a charitable lens. This reinforces both Hussain et al. (2002) and my earlier study (Rizvi, 2015) highlighting parental fears about the lack of acceptance of their disabled child within Pakistani society. This finding strongly refutes the UNICEF (2003) report, which suggested that although inclusive education had not been developed extensively as an ideology within Pakistan, the public generally held positive attitudes towards disabled individuals. The reportedly positive attitudes of the Pakistani public may be because disabled children have traditionally been educated by religious institutions which, despite enforcing gender segregated schooling, strongly focus on equal access for all learners (UNICEF, 2003). However, disability is still perceived through a charitable lens within Pakistan as many charities and NGOs operate special schools. Interestingly, special education falls under the joint responsibility of the Ministry of Education and the Ministry of Social Welfare and Special Education in Pakistan.

With British-born mothers, there was never any consideration of living elsewhere. They had been born and raised in this country and had known the UK as their only home, and hence, they held no feelings of any regret or disappointment about Pakistan. Their opinions about Pakistan had been formed based on their interactions with their relatives in Pakistan, their parents' experiences of living there before migrating, Pakistani TV dramas and news, and to a lesser extent their short trips to Pakistan when they were young. Interestingly, they held an even more alienated and pessimistic picture of Pakistan than the mothers who had been raised there. Much of their perceptions had been built on what Kiran termed *"horror stories"*, relayed through the media or acquaintances about how disabled individuals were simply locked up in their rooms. Although these narratives were not fictitious and it is true that from time-to-time the media uncovers stories of how a disabled individual was being treated in a cruel and inhumane manner, it is certainly not the story of all disabled individuals in Pakistan and nor is it limited only to a Pakistani context. Yet it for British-born mothers, this partial truth forms their whole truth and perception of Pakistan.

In Pakistan, you know you'll never have access to all the help and support that you get here [Britain] … you hear all these horror stories … special needs children don't get anything … they don't have specialist schools or anything like that, do they? They're locked up in their rooms and all these horror stories that you hear … I just think we're very lucky that we're living in Britain and for Ahmed to have all these facilities and that we're supported well.—Kiran

Although I was born here, my family and my parents are from Pakistan, so are my in-laws, so their understanding might be different. I do know that in Pakistan, especially in villages what happens is that these disabled children, they're hidden away … in Sialkot [Pakistani city], there is a special needs provision but I think it's more for them who can afford it … Yeah, education in Pakistan, even in the government schools you've got to buy your books and uniforms and whatever. You don't have all that here [Britain], books and education is free until the age of 16 or 18 … maybe because I'm born and bred here, the doctors tell you everything and try and do tests, whereas in Pakistan they don't have this kind of system, if you're ill you go see a doctor but if you're not better by the evening you go to another one … If I was to go to Pakistan and if God forbid something happened, I wouldn't know where to go … it's just a money-making scam, be it a doctor or be it anybody else.—Maria

Like the first-generation migrant mothers in my study, the British-born mothers also discussed the privileges of living in a first-world country, and more importantly, of feeling fortunate to have avoided raising their disabled child in Pakistan. At times, it seemed that their perception of living in Pakistan was a consequence of internalising how developing countries are often portrayed in the media and in government discourse; being scammed by doctors, late and inaccurate diagnosis, a lack of emphasis on education, institutionalising disabled people or keeping them locked up at home, and the lack of any professionalism in general. It is interesting to consider that Maria's experiences of the incorrect and late diagnosis of her son, Aamir's disability did not occur in Pakistan; it happened in the UK and yet her perception of Pakistan leads her to believe that things could have been far worse in Pakistan. However, it is important to note that mothers in this research did have grievances with the way the British state had served their communities as well as how well the state had supported them in exercising their citizenship, which I now turn my attention to.

SHIKWA WITH ENGLISTAN

There are no clear-cut distinctions between British-born mothers and migrant mothers that would suggest that one group is passive whereas the other group is active in terms of exercising their citizenship. Maternal narratives were intricately tied to who felt ready to claim ownership of 'Englistan' as their home. Some mothers arrived early, literally and metaphorically, at a point where they felt a sufficient sense of belongingness to the UK to express grievances towards their homeland. Other mothers were still in the process of learning to let go of the fear that comes with a temporary legal status, to acquire a sense of membership and security that comes with the legal status of citizenship.

> On this little island
> Where we're all surviving
> Politeness mixed with violence
> This is England …
> Is Britain great? Well hey don't ask me
> But it's where I live and why my heart beats. (Riz Ahmed "*Englistan*" cited in Westcott, 2016)

For first-generation migrant mothers like Parveen, there was still this perception that the provisions and services provided to their disabled children by the British state was somehow too much compared to the number of years they had spent in the UK, which had to be repaid in ways other than through their tax contributions.

> When Sehr was diagnosed, I saw all these professionals in the meeting, so being worried I asked the, do we have to pay all of you? I was from Pakistan, how would I know how the UK system works? I hadn't had any kids at hospital before. I think if I sold my house ten times over, it still wouldn't be enough to pay [medical expenses]. I wouldn't have been able to provide the care my children received here [Britain]. I feel obligated to do something for this country because it's supported my daughters. I feel I need to do something in return for the community here. I've started volunteering in elderly care homes, cooking for them once a week. I feel even doing this little thing makes me feel good. I want to do something for the community here … During last year's flooding, I raised £10,000 with my friends and went to flooded areas to help …—Parveen

This was a very compelling discussion because it reveals how migrants view their relationship with the welfare state. Whereas British-born citizens expect to contribute taxes proportionate to their income because they recognise that is how you pay for state services, migrants might view their tax contributions as recompense for the social and legal freedoms, law and order, and good governance that they enjoy in this country in addition to the right to an education, healthcare and welfare support that the UK provides its citizens. With Parveen, it is not that she was unfamiliar with how the UK welfare state operates, but rather she felt compelled to go above and beyond the expectations of what a citizen should do—such as paying taxes—since she depended so heavily upon the welfare state for her child's needs. The financial cost of each medical operation and the quality of the extended care provided to her daughters deeply weighed on Parveen's sense of obligation as a citizen. She not only wanted to tread carefully by checking with medical professionals if she was expected to pay, but upon learning that she did not have to pay, she made a conscious decision to give back to British society in the form of voluntary service in elderly care homes. Parveen's rationale for her community outreach work was also embedded in her Islamic ethos, and how she believes that British Muslims must represent themselves to wider British society. It is difficult to say whether she had internalised the problematic depiction of British

Muslim communities as a burden, feeling that she had to 'fix' this negative stereotype. She also felt that her tax contributions were not enough to repay British society for what the welfare state had provided for her daughters, even if she sold her house "*ten times over*".

Tahira, Maham and Alina's narratives also displayed faith that state institutions would protect the rights of their disabled children. They each believed that the UK system was not only governed by specific laws that protected their disabled children, but also that it did not hold their foreign heritage as a reason to block the provisions and services that their disabled children needed and were entitled to. This does not mean that first-generation migrant mothers did not express any dissatisfaction, or that they had no dissenting opinions about how the state should be involved with regards to provisions for their disabled child.

> I feel professionals are sometimes unable to understand us. There's a big communication gap … for instance, two to three weeks back, I voluntarily participated in [hospital] feedback interviews regarding Sehr and Amber, but [when I attended] they were surprised to see! … They rightly emphasise the patient but only as an individual, they don't see the bigger picture and that's where the problem starts. It's not their problem that because of Sehr's death, my Suroor [Parveen's middle daughter] completely blanked out in depression, but this isn't their problem. Why would they consider Suroor? So then you and your children begin to struggle but you're unable to explain this to them.—Parveen

All the maternal narratives referenced a communication gap and some level of mistrust with professionals, which stemmed from their experiences with professionals who did not consider or were cognisant of their family's needs. Mothers acknowledged that professionals focused on their disabled child but they did so in isolation, which missed the bigger picture. For instance, Parveen discussed how medical professionals could have been improved their services by using a holistic approach in the aftermath of Sehr's death, pointing out that her middle daughter, Suroor, fell into depression as a result of the trauma. However, her resistance was more cautious in nature, as she questioned whether it was the medical professional's responsibility to consider holistic needs, or how bereavement affects a patient's family members, and whether any support should have been set in place for Suroor. In Maria's case, she was pressured by both medical and school professionals to agree to have a PEG fitted in Aamir's stomach so that feeding time during lunchtimes at school could be quicker and easier. However, they completely overlooked the fact that for Maria's family, the evening meal was the one activity that they could still share with Aamir as a family. Her reluctance to submit to such professional pressure could be perceived as maternal denial about the interventions her child needs, but it also reflects how professionals often overlook a disabled family's holistic needs.

For British-born mothers like Maria, Shehnaz, Kiran and Saira, their grievances were far more grounded in the belief that they were well within their rights as citizens to question and express grievances about state services their families' received, without fear of being perceived as ungrateful. For instance, Shehnaz questioned the local council's delayed placement decisions for her daughter, Amna; or when Saira complained to her family intervention worker after her daughter, Farha's school made an inaccurate (and in Saira's opinion, racist) allegation that domestic abuse was responsible for Farha's depression when it was due to school bullying; or when Kiran raised the issue of how certain provisions had been left out from her son, Ahmed's annual review; in all these situations, these British-born mothers were more aware of the presence of institutional gatekeeping in the allocation of resources which was mediated by a familiarity with the British system. It is a familiarity afforded them through multiple generations of engagement with British institutions, a familiarity that first-generation migrant mothers did not possess.

McMillan Cottom (2019) suggests that for minoritised communities the disenfranchisement, racism and exclusion are generational; however, each generation inherits the knowledge of how to interact with the gatekeepers of resources from its previous generation. With British-born mothers, their knowledge of how to deal with discrimination and racism within institutional bureaucracy was partly learnt from their experiences of going through the English schooling system themselves, but also inherited from their parents and grandparents.

> Consequently, my grandmother and mother had a particular set of social resources that helped us navigate mostly white bureaucracies to our benefit … I remember my mother taking a next-door neighbor down to the social service agency. The elderly woman had been denied benefits to care for the granddaughter she was raising. Her denial had come in the genteel bureaucratic way—lots of waiting, forms, and deadlines she could not quite navigate. I watched my mother put on her best Diana Ross Mahogany out-

fit: a camel-colored cape with matching slacks and knee-high boots … It took half a day, but something about my mother's performance of respectable black person—her Queen's English, her Mahogany outfit, her straight bob and pearl earrings—got done what the elderly lady next door had not been able to get done in over a year. I learned, watching my mother, that there was a price we had to pay to signal to gatekeepers that we were worthy of engaging. It meant dressing well and speaking well. It might not work. It likely wouldn't work, but on the off chance that it would, you had to try … There is no evidence of access denied. Who knows what I was not granted for not enacting the right status behaviors or symbols at the right time for an agreeable authority? Respectability rewards are a crapshoot, but we do what we can within the limits of the constraints imposed by a complex set of structural and social interactions designed to limit access to status, wealth, and power. (McMillan Cottom, 2019, p. 89/90)

McMillan Cottom (2019) refers to the relationship of Black American communities with the state, and how countless decades of state-led criminalisation and marginalisation of their communities has meant that Black Americans continue to suffer racial disparities in housing, education, employment, health and other sectors. She relays her own experiences of being raised in a household where her mother and grandmother both knew the politics of code-switching to how White people talk, and how this could unlock institutional gatekeeping. It is important to note here that McMillan Cottom acknowledges that there is no guarantee that this generational knowledge works and that it might be a gamble, but she also posits that not engaging in code-switching may give the impression that Black American families are either not intelligent enough to articulate their needs or are not worthy of engaging with. I find her reflections powerful because they capture the generational trauma and the resulting knowledge systems that are developed to deal with various forms of institutional racism and disenfranchisement. Her accounts closely resonate with the experiences of British-born and first-generation migrant British Pakistani mothers in the current study.

First-generation migrant mothers, unless they were part of a supportive joint-family system like Maham, had limited access to generational knowledge of the system in this country needed to access support that was rightfully theirs. They also had to walk the line very carefully because they were all too aware of how they would be perceived—as foreign-born immigrants not British-born citizens. They could be fortunate in being able to access informal support systems such as local community centres who could help them navigate the UK special education system; however, in this research, the community centre itself had become a gatekeeper to accessing support for a few families whom they disapproved of. British-born mothers were positioned as more knowledgeable of the British special education needs system compared to their first-generation migrant peers.

> ... they're born in Pakistan and they've obviously come over here after their marriages ... a few of them have approached me for some advice, to fill in forms, to speak to me about things because they see me with my children and they think, well she's done well, we can go to her ... it's definitely an advantage to be born here. I don't depend on anybody, I don't have to wait for someone to do things for me, I can just go and crack on with it.—Shehnaz

British-born mothers in this research considered themselves to be better equipped in terms of resources than first-generation migrant mothers, because they were more familiar with the British schooling system, values and other wider institutions. This is similar to the study by Heer et al. (2012) which found that British-born second-generation South Asians were more familiar with British culture and had greater engagement with educational and medical providers, and were therefore more likely to relate to Western conceptualisations or models of disability.

For British-born mothers, there is also a generational 'financial stake' in the system in that their parents and grandparents have paid into the system. However, these British-born mothers still felt on the margins when it came to accessing services and provisions, which engendered a stronger sense of disappointment that comes with having generational ties to Britain and yet still feeling like an outsider. It is also important to note that British-born mothers were still navigating the special education system that was unfamiliar to their parents, but because they had a greater understanding of how UK institutions and laws work they were more likely to realise when their children were let down by the system.

> It's been a couple of years now ... I mean, if I knew at that time that his Statement meant [that] even as a special needs child he's [Faraz] entitled to an education in a mainstream school ... if you've got a Statement ... you can apply to have an education ... I would've done things differently. But I didn't have that knowledge, nor was it explained to me ... whenever we had meetings, there was always someone [present] from the educational system

who was attached to the school or from the council or whatever depart-ment, they should've explained to me properly what my child was entitled to ... you see with Faraz I didn't have a family interventions worker so I didn't really have that support there, whereas now I have that support [with Farha and Zara] ... I've got someone there to fight my corner and make me aware that, 'hold on a minute the school, the educational system can't do this' ... I'm just thinking that if I didn't have that family interventions worker and if she hadn't contacted the school then I'd just be going nuts about Zara, trying to sort out stuff with her school ... you know, you're made to feel like a bad parent then.—Saira

Despite being born and raised in Britain and being familiar with the wider schooling system, nonetheless, Saira had to pay a personal 'cost' with the permanent exclusion of her eldest son, Faraz in order to gain that knowledge to enable her to be taken seriously by educational professionals with her daughters, Farha and Zara. No one from the local authority had advised her of her child's rights, and so her interactions with these profes-sionals had led her to believe her own mothering skills were at fault and that she was simply not trying hard enough with Faraz. It was only through self-learning that she learnt to involve a family intervention worker who could fight her corner. One wonders whether this family intervention worker was considered 'worthy of engagement' by educational profession-als when it came to Farha and Zara, in contrast to Saira, and if Saira alone could have provided the *"right status behaviors or symbols"* (p. 90) which McMillan Cottom (2019) suggests that professionals look for in their interactions. Also, in Saira's narrative is her realisation that the education system could not treat her children like lesser citizens, and that they could not hold back provisions from her children which they were entitled to by law.

It is that certainty and confidence which Saira possesses by virtue of being British-born and bred, as well as through paying the personal price with Faraz, that she now understands her rights and her children's entitle-ments. With Parveen, Tahira and Alina as first-generation migrants, their confidence to ask questions of the system and not fear coming across as ungrateful subjects was still mediated by their newness to the system. It is possible that in the future, their children gain the epistemic confidence and certainty that Saira, Kiran, Shehnaz and Maria developed as second-generation British citizens from their experiences of being raised within the British system.

In discussing the maternal narratives within this chapter, it has become clear that regardless of immigration trajectories, both first-generation migrant mothers and British-born mothers in this study may be subsumed into the broader discourse about how ethnic minoritised communities experience inclusion and exclusion within Britain's special education needs systems. Unfortunately, the current political emphasis on stricter immigration controls and cultural assimilation is pressuring ethnic minority communities to prove that they 'deserve' to be in the UK even if this country is the only place they have ever known (Mulvey, 2010).

REFERENCES

Bhatti, G. (1999). *Asian children at home and at school: An ethnographic study.* Routledge.

Cottom, T. M. (2019). *Thick: And other essays.* The New Press.

Dahya, B. (1974). Pakistani ethnicity in industrial cities in Britain. In A. Cohen (Ed.), *Urban ethnicity* (pp. 77–118). Tavistock.

Goodley, D., Runswick-Cole, K., & Mahanoud, U. (2013). Disablism and diaspora: British Pakistani families and disabled children. *Review of Disability Studies: An International Journal, 9*(2 & 3), 63–68.

Grierson, J. (2018, August 27). Hostile environment: Anatomy of a policy disaster. *The Guardian.* Retrieved December 15, 2020, from https://www.theguardian.com/uk-news/2018/aug/27/hostile-environment-anatomy-of-a-policy-disaster

Hafeez, A. (2020). Special education in Pakistan: A critical analysis. *A Journal of National School of Public Policy, 41,* 161–182.

Heer, K., Rose, J., & Larkin, M. (2012). Understanding the experiences and needs of south Asian families caring for a child with learning disabilities in the United Kingdom: An experiential-contextual framework. *Disability & Society, 27*(7), 949–963.

Hussain, Y., Atkin, K., & Ahmad, W. (2002). *South Asian disabled young people and their families.* Joseph Rowntree Foundation, The Policy Press.

Meekosha, H., & Dowse, L. (1997). Enabling citizenship: Gender, disability and citizenship in Australia. *Feminist Review, 57*(1), 49–72.

Meekosha, H., & Soldatic, K. (2011). Human rights and the global south: The case of disability. *Third World Quarterly, 32*(8), 1383–1397.

Miles, S., & Singal, N. (2010). The education for all and inclusive education debate: Conflict, contradiction or opportunity? *International Journal of Inclusive Education, 14*(1), 1–15.

Mulvey, G. (2010). When policy creates politics: The problematizing of immigration and the consequences for refugee integration in the UK. *Journal of Refugee Studies, 23*(4), 437–462.

Population Census. (2011). *Pakistan bureau of statistics.* Government of Pakistan. Retrieved December 15, 2020, from http://www.pbs.gov.pk/content/population-census

Rizvi, S. (2015). Exploring British Pakistani mothers' perception of their child with disability: Insights from a UK context. *Journal of Research in Special Educational Needs.* https://doi.org/10.1111/1471-3802.12111

Sayad, A. (2004). *The suffering of the immigrant.* Polity Press.

UNICEF. (2003). *Examples of inclusive education Pakistan.* UNICEF Regional Office for South Asia.

Westcott, L. (2016, July 13). Riz Ahmed talks HBO's 'the night of' and the meaning of 'Englistan'. *Newsweek.* Retrieved December 15, 2020, from https://www.newsweek.com/riz-ahmed-night-englistan-479967

Williams, W. (2020). Windrush lessons learned review. The APS Group on behalf of the Controller of Her Majesty's Stationery Office. Retrieved December 15, 2020, from https://www.gov.uk/government/publications/windrush-lessons-learned-review

CHAPTER 7

Concluding Thoughts …

I have written this book at a time when one recent study has suggested that British White homeowners move out of areas where the in-coming homeowners have Pakistani—primarily Muslim—sounding names, regardless of whether these areas have high or low non-White ethnic populations (Easton & Pryce, 2019). Additionally, when a recent survey by YouGov (2018), which examined the contributions to the UK of immigrants from different countries, found that Indian immigrants were evaluated positively whereas Bangladeshis and Pakistanis—who are overwhelmingly Muslim—were adjudged as having made negative contributions to British society. It is also written at a time when studies suggest that the Covid-19 pandemic has hit minority communities disproportionately in terms of health and financial implications (Haque et al., 2020). They suggest that not only are members of minoritised communities more likely to be in public-facing jobs than the White community, but they are also more likely to have to depend on their savings for their everyday expenditures and to fulfil their basic needs such as food, rent or utility bills, to need to borrow money from family and friends to make ends meet, or to resort to missing meals (or missing meals more often) as a result of their financial difficulties (Haque et al., 2020).

In light of all this, it is infeasible for the field of special education and disability studies to continue to research and understand the experiences of minoritised disabled families, without considering the full implications of the wider social issues and inequalities that intersect with their everyday

S. Rizvi, *Undoing Whiteness in Disability Studies*,
https://doi.org/10.1007/978-3-030-79573-3_7

experiences. By centring on the experiences of minority mothers, I am arguing for a more thoughtful and critical engagement by professionals who interact with minoritised disabled families, asking educational professionals and academics to critically evaluate their own assumptions about inclusive practices and consequently to enable a more positive relationship with these families. Current professional relationships with minority communities rely heavily on a narrowly defined and problematic conceptualisation of 'inclusion', 'partnership', 'advocacy', 'South Asian', 'Muslim', 'mothering' and 'disability'. This book pushes for more accountability and ethics of care in how services and provisions are set out for minoritised disabled families generally, and British Pakistani disabled families in particular.

As I have argued in this book, the mothering role entails a great deal of caregiving and other everyday duties which are not considered relevant to education discourse. These imposed boundaries have been challenged and are gradually being dismantled by mother-scholars such as Landsman (2005), Runswick-Cole and Ryan (2019), Green (2007), and Rogers (2011). However, the experiences of ethnic minority mothers of disabled children have remained invisible.

BARRIERS

This book highlighted a number of barriers that minoritised mothers experience at different levels, which invisiblise their contributions, experiences and the labour involved in mothering their disabled children.

Community Level

Formal service providers often lack the cultural literacy and familiarity needed to engage with minoritised communities. Community centres such as Anokha, which arranged my introduction to some of my participants, fill this crucial gap. These community centres provide vital services in the form of respite care, signposting of relevant services, holding workshops for families and carers, arranging family oriented religious and sociocultural events, and advocating for the social and educational needs of disabled children. Anokha undeniably provides these valuable services to its local community, however, the maternal narratives in Chap. 5 revealed how Anokha needed to transform its organisational processes so that it could provide the fullest protection and advocacy services for mothers from these communities. Over a period of time, any organisation runs the risk of becoming hierarchal in nature, developing several gate-keeping layers that might hinder access for mothers needing support and ultimately preventing it from creating any change. It is therefore imperative that community centres develop their own 'best practice' standards of transparency, accountability, ethics of care and confidentiality that are strictly followed when dealing with families with whom they share ethnic, familial and generational ties. It is also essential that these standards are not breached or disregarded in favour of cultural values that support patriarchal family and community structures.

Religious Level

Despite the fact that Islam was perceived by the mothers in my study as an inclusive and positive influence, their experiences within mosques and other religious spaces were not (Fatima, 2016). This was a considerable barrier for mothers because religion plays a major role in all their lives. This study found that certain aspects of cultural patriarchy were often presented as religious edicts which have significant implications in how disability is perceived by the community, how caregiving becomes gendered and limited to a domestic sphere, how disabled Muslim families are restricted from fully participating and exercising their rights as Muslims, and how disabled families can miss out on informal support due to the ableist and patriarchal nature of informal support structures. Religious institutions therefore require major reform in how they can become more inclusive and can utilise their influence to instil inclusive and egalitarian values within the wider Muslim community.

School Level

Some mothers in this study were failed by both mainstream and special schools, who did not value their maternal ways of working and which consequently affected their child's inclusion. The biggest barrier that mothers faced was that the responsibility for their child's inclusion was not shared by schools. There was an assumption from schools that all mothers would somehow inherently 'know' how to navigate the SEN system the first-time around and also that they would unquestioningly go along with every school decision. We observed this in the way Parveen's daughter, Amber was included within a mainstream school without considering whether or not her needs could be supported in the school, as well as how they conducted their relationship with Parveen. This presumptuous attitude from professionals that minoritised mothers were merely passive stakeholders requires dismantling, especially when there are language barriers present as we noted in the case of Tahira, which further invisibilised her efforts.

Another barrier that schools erected before minoritised mothers was the lack of signposting of appropriate support services and advocacy groups to inform them of their children's entitlements. Whilst all mothers wanted their child's overall outcome to be the school's main focus, nonetheless, they held certain expectations that the school would signpost them towards appropriate services and many mothers were dissatisfied with the information they received about their child's entitlements, as we saw in Saira's case. This suggests that many mothers experience an unequal and unbalanced parent-professional partnership as a result of schools holding back information that is necessary for maternal decision-making, which ultimately undermines the ability of mothers to participate fully within their child's education. The consequences of withholding information of their children's rights and entitlements from minoritised parents are grave, with potentially life-threatening impacts on disabled children.

For instance, a series of fixed term as well as permanent exclusions led Saira's son, Faraz to become severely depressed and ultimately develop suicidal thoughts. His exclusion was brought on by a multitude of factors such as his school labelling him 'disruptive' rather than acknowledging the fact that he was being bullied at school, that the school felt they were ill-equipped to support his behavioural needs and subsequently 'writing him off' due to his ADHD, and because the school and local authority completely disregarded the 'pupil voice' of Faraz that preferred mainstream school which was contrary to the underlying principles of the SEND Code of Practice (Department of Education and Department of Health, 2015). Saira very fortunately was able to prevent what could have been the tragic suicide of her son. Nonetheless, she blamed the entire situation on institutional failure. Unfortunately, Faraz is just one of many racialised and minoritised children who have been, and continue to be let down by the education system. In extremis, we can see the result of such institutional negligence in the tragic murder of Tashaun Aird, a 15-year-old boy who was killed in Hackney, east London in May 2019 by teenage gang members who wrongly assumed he was a member of an opposing gang. Tashaun had been permanently excluded by his mainstream school and referred to 'alternative provisions', without any consideration of whether his permanent exclusion may impair his safety, wellbeing and overall educational outcome (Berg, 2020).

The reality is that too many children like Faraz and Tahira's son, Farrukh are persuaded by professionals that they are unprepared and ill-equipped for mainstream school, further excluding them and potentially adversely affecting their mental health (Done & Knowler, 2020). In a system where racial inequalities can already be witnessed across different sectors and systems, ignoring the rights of minoritised disabled children leads schools to become active sites where exclusion and inequity are perpetuated.

Local Authority Level

A major barrier that local authorities present to minoritised mothers is the lack of transparency in the inner workings of the school admissions process. There is a presumption that all parents are familiar with the placement admissions process for their disabled children. Despite the fact that the SEND Code of Practice (Department of Education and Department of Health, 2015) promotes *"the participation of children, their parents and young people in decision-making ... [and] greater choice and control for young people and parents over support"* (p. 19), we saw in Shehnaz's narrative that the process is time-consuming and requires mothers to be informed about the available placement options, and does not always consider parental and pupil choice. Minoritised mothers like Shehnaz experience exclusion during this process on multiple fronts; for instance, a delay in the admissions process for her daughter, Amna meant that Shehnaz had to expend all her energies on gathering evidence for submission to the local authority in support of Amna's case, which is time that could have been spent supporting and caring for Amna, Tariq and her other children.

RECOMMENDATIONS

This book has reported how institutional barriers can impede minoritised mothers in performing their mothering duties for their disabled children. Arising directly from the findings of the current study, the following recommendations are an illustration of how different formal and informal spaces can become inclusive of minoritised mothers with disabled children.

Greater Representation of Mothers in Religious Settings

One key feature that has been highlighted in this book is the way in which religion can positively shape the experiences of minoritised mothers of disabled children. In talking to the mothers in this research, I noted how religion and religious spaces could offer informal support to disabled families in terms of respite care, the opportunity to practice and participate in religious activities and events for disabled children and young people, and to promote inclusive values within the wider British Muslim community. There is a vital service gap that mosques and religious centres can fulfil in terms of offering communal child-rearing support and respite care for disabled children, which would lessen its emphasis as a maternal duty and so make caregiving less gendered; moreover, this community-wide support would enable mothers to exercise self-care and feel part of a community that can actually offer them some tangible assistance. This is neither farfetched nor a new idea. Many studies with British South Asians in various disciplines have highlighted how, rather than being problematized by official discourse, religion should instead be viewed as a valuable untapped resource that can bridge the service gap between the welfare state and minoritised communities in this country. A recent joint initiative, *#WomenInMosquesConversationToolkit* between the Muslim Council Britain (MCB) and the Islam-UK Centre at the University of Cardiff (2020), as well as the *Women in Mosques Development Programme* also by the MCB (2018), are excellent examples of how British Muslim organisations have begun to devise ways of supporting Muslim women to take leadership and governance roles in the running of mosques in the UK. Whilst these initiatives do not specifically focus on Muslim disabled families, they do nonetheless enable Muslim women from different communities to decide on how to make these religious spaces become more accessible, constructive and family friendly and in the process less patriarchal.

Mosques can also organise workshops that challenge cultural stereotypes surrounding theological explanations of disability, the religious obligations of individual disabled adults and children, and the gendered nature of caring. These workshops can be facilitated by inviting religious scholars to discuss and promote positive inclusive attitudes around the families of disabled children, reducing misconceptions and stigma and removing community barriers towards their wider religious and community participation. This process can also be strengthened by inviting speakers who are professionals in the fields of health, education and social care and, importantly, who are from the local community and so possess an insider status that will enable them to understand the community's religio-cultural sensitivities.

Cultural and Community Centres

There is little doubt about the important work that is performed within these informal spaces, often with limited funding and dependent on the physical and emotional labour of community volunteers. It is incumbent on the government to offer sufficient funding to these centres, which have the potential to be a vital bridge of trust and communication between formal government institutions and minoritised communities. A recent report by the UK Women's Budget Group (2019) informs us that since 2010, the Local Government Association (LGA) has experienced £16 billion of funding cuts from council budgets in England, and the National Council for Voluntary Organisations (NCVO) has similarly reported that over 50 per cent of councils have made cuts to voluntary sector initiatives and schemes. The consequences of these austerity policies is that services that are often a vital lifeline for minoritised communities such as social care, housing services and welfare assistance as well as services that are classified as ancillary such as English to Speakers of Other Languages (ESOL) courses for adults, have all faced reductions in funding within councils in some of the most deprived areas of Britain. This has disproportionately affected Black and South Asian women in this country (UK Women's Budget Group, 2019).

It is vital that more community centres such as Anokha are set up that can provide a real choice to ethnic minority communities, offering a safe informal space for disabled families to build support networks. More importantly, it is critical that community centres such as Anokha develop their own cultural competency training and codes of practice that dismantle gatekeeping and overly bureaucratic and patriarchal practices, so that they can better safeguard and support the most vulnerable members of the community. There is also an opportunity here for these informal spaces to collaborate with mosques and interfaith centres, so that they can pool their resources to change community-wide misperceptions of disability. For instance, the work over the last two decades of the Texas Industrial Areas Foundation, an interfaith and multiracial forum of community organisers in the USA, is a brilliant example of how religious networks can be utilised to rebuild and empower communities to realise their democratic rights (Collins & Bilge, 2017). Using an intersectional lens as an analytical tool, these informal spaces have provided entry points to members from different communities to participate without fear of cultural or religious policing.

Deconstructing School Partnerships

In Chap. 2, mothers in this study discussed the frequently toxic nature of their relationships with their disabled children's schools, which in turn throws up a number of valid questions that require our attention. Why did Farrukh's school need to view a home video before they would believe Tahira that Farrukh happily eats Pakistani food at home? How much of Tahira's already restricted time as well as physical and emotional energy was wasted in this effort? Why did Parveen face such a hostile attitude from Amber's school? Why did Imran's school expect that Alina would endorse her son's isolation at school? Why did Aamir's school believe it was fine to pressure Maria into consenting to have her son fitted with a PEG? Why was Farha's depression wrongly and all-too-easily attributed to a suspicion of domestic violence within Saira's home?

In each of these instances, the schools lost maternal trust and respect, both of which are key to any authentic collaboration. The traditional ways of working with minority families must be re-examined because not only are they based on a flawed assumption that the child-rearing strategies and knowledge of minoritised mothers are deficient and out-dated, but they also fail to recognise the intersectionalities of these mothers and result in partnership models that further silence and invisibilise minority mothers (Harry, 2008). Home-school collaboration should reimagine "*schools as communities and to the communities of schools*" (Harry, 2008, p. 384). This should involve hiring more BIPoC teachers, training existing school staff in cultural and racial literacy, rethinking how curriculum and school support can be redrafted to include community and familial funds of knowledge from minoritised communities (Yosso, 2005), and rethinking the appropriateness of the current assessment and diagnostic tools that assess disability in children and young people from Black and ethnic minority backgrounds. Calderon-Berumen (2019) suggests that the only way for schools to let go of the deficit model is by bringing the "*cultural curriculum of the home in school settings, even when this implies the inclusion of another or multiple languages, will include accepting different ways of being, doing, and knowing. Thus, it will enhance everyone's learning process*"

(p. 11). I feel her position offers schools an entry point to reducing the existing power dynamic and celebrating the communities in which they are situated.

Changing the Institutional Lens

It is a challenging task to change how local authorities and official institutions—that are predominantly White and middle-class—view and work with minoritised families, unless we push for radical change. Collins and Bilge (2017) suggest that policymakers can develop and implement public policies that reduce social inequalities if they ascribe to an intersectional lens, one that examines the root causes of discrimination based on religion, race, ethnicity, sexual orientation and other social categories; a good example is the Charter of Fundamental Rights of the European Union (EU). However, following the UK's exit from the EU, the Charter of Fundamental Rights has ceased to have effect in this country and the danger is that human rights will become a limited single focus issue. Ultimately, this could lead to an institutional neglect of the multiple layers of intersecting oppressive systems that minoritised communities face and that limit their lived experiences on a daily basis.

There are various solutions that can be implemented to pre-empt and mitigate this course. For instance, institutions can increase the representation from minoritised communities within their formal governing boards, and ensure that policymakers are acutely aware of the barriers that these communities face through mandatory cultural and racial literacy training. Another solution is that the UK needs to develop its own Bill of Rights which mirrors the EU's Charter of Fundamental Rights that is designed to prevent discrimination based on religion, race, ethnicity, sexual orientation and other social categories, that gives room to challenge those UK policies which specifically target minority communities, that prevents enslavement, and that protects migrants and asylum seekers from being removed from the UK.

Charities and grassroots organisations also have a crucial role to play here, by providing greater external oversight in holding official institutions to account for their decisions. A case in point and an excellent example of how official governing bodies can be held accountable, is the recent legal action by the Runnymede Trust and the Good Law Project against the UK government with regards to its closed recruitment process for the post of Head of NHS Test and Trace (Begum, 2020). They jointly argued that the government breached the Equality Act 2010 in not advertising these executive level NHS posts, with implications of indirect discrimination against minority communities who are already notably underrepresented in top management jobs in the NHS. Another outstanding example of how local organisations can push for greater institutional accountability is the MCB's Centre for Media Monitoring; they have continuously called out UK media companies in relation to media discourse of minoritised communities, specifically those outlets involved in publishing over 100 blatantly anti-Pakistani and anti-Muslim stories.

THE FIRST STEP TO UNDOING WHITENESS

We must keep the perspective that people are experts on their own lives. There are certainly aspects of the outside world of which they may not be aware, but they can be the only authentic chroniclers of their own experience. We must not be too quick to deny their interpretations or accuse them of "false consciousness". We must believe that people are rational beings, and therefore always act rationally. We may not understand their rationales, but that in no way militates against the existence of these rationales or reduces our responsibility to attempt to apprehend them. And finally, we must learn to be vulnerable enough to allow our world to turn upside down in order to allow the realities of others to edge themselves into our consciousness. (Delpit, 1988, p. 297)

In looking to plot a way forward, how can we validate and celebrate different ways of knowing without viewing them as threats to our existing institutions? At the start of this book I suggested that our knowledge of South Asian disabled children and their families has been reduced to cultural and language issues, which prevents them from improving their overall experiences. Therefore, this leaves the onus of their poor experiences resting on these communities themselves. I concur with Delpit in that the problem is down to institutional culture, rather than the culture of minoritised families. Delpit (1988) asked educational practitioners nearly four decades ago to think about the role of cultural power and how it creates othering. She posited that the cultural framework of the USA is such that it has, since its beginning, created the belief that the culture of Black Americans is inferior to White Americans. What this damaging narrative has done at an epistemic level is to create a level of comfort and a flawed presumption that the education system does not discriminate according to race, gender, sexual orientation, religion or socioeconomic status, but rather delivers for everyone. By this rationale, the responsibility for poor educational outcomes can be laid squarely at an individual or familial lack of educational aspiration or motivation to succeed, rather than due to structural or systemic failures, barriers and gaps in the American education system. This also implies that any change can only come from the bottom-up, as opposed to top-down, thereby abdicating all accountability and responsibility on the wider education system to provide equity and equality of opportunity for all children and young people in the USA (Delpit, 1988).

Similar master narratives can also be observed when we look at UK schools and their contentious relationships with minorities communities. The maternal accounts in this book speak truth about how the special education system in this country does not see their children in their entirety, as individuals who are rooted in their own culture, religion, ethnic heritage and other inter-categorical positionings. In some cases, the differences of these minoritised families were perceived as threatening to normative ways of working and as such were devalued by educational, medical and social care professionals who did not view the mothers as experts or as having the intellect to articulate their own experiences. This means that the services and provisions available in this country will always be developed with only a partial understanding of what minoritised disabled families need. Believing that intersectional histories are irrelevant to educational issues will only worsen the experiences of these families. Some schools and organisations are already doing the good work of bridging this mother-school divide, using their power to transform the experiences of minoritised communities whilst also letting minoritised communities transform their own organisations for the better. There is an opportunity here for schools and other institutions to understand how inclusion is conceptualised by minoritised disabled families—these families strive for inclusion on a more fundamental level, one that acknowledges the consequences of multiple axes of exclusion and discrimination and that pushes for the acceptance of their child's rights to an education and to all available opportunities. Inclusion for minoritised disabled families has to be intersectional, and at an epistemic level must move beyond the narrowly defined locational inclusion that is often discussed when we examine the educational experiences of disabled children and young people.

REFERENCES

Begum, H. (2020, November 16). Why Runnymede is suing the government over its Covid-19 hiring practices. *Runnymede Trust*. Retrieved December 15, 2020, from https://www.runnymedetrust.org/blog/why-runnymede-is-suing-the-government-over-its-covid-19-hiring-practices

Berg, S. (2020, December 22). Tashaun aird: Family of murdered boy critical of school exclusion. *BBC News*. Retrieved December 23, 2020, from https://www.bbc.co.uk/news/uk-england-london-55353958

Calderon-Berumen, F. (2019). Resisting assimilation to the melting pot. *Journal of Culture and Values in Education, 2*(1), 81–95.

Centre for Media Monitoring. Retrieved December 15, 2020, from. https://cfmm.org.uk/

Collins, P. H., & Bilge, S. (2017). *Intersectionality*. Polity Press.

Delpit, L. (1988). The silenced dialogue: Power and pedagogy in educating other people's children. *Harvard Educational Review, 58*(3), 280–299.

Department of Education and Department of Health. (2015). Special educational needs and disability code of practice: 0 to 25 years. Retrieved December 15, 2020, from https://www.gov.uk/government/uploads/system/uploads/attachment_data/file/398815/SEND_Code_of_Practice_January_2015.pdf

Done, E. J., & Knowler, H. (2020). A tension between rationalities: "Off-rolling" as gaming and the implications for head teachers and the inclusion agenda. *Educational Review*. https://doi.org/10.1080/00131911.2020.1806785

Easton, S., & Pryce, G. B. (2019). *Not so welcome here? Modelling the impact of ethnic in-movers on the length of stay of home-owners in micro-neighbourhoods*. https://doi.org/10.1177/0042098018822615.

Fatima, S. (2016). Striving for God's attention: Gendered spaces and piety. *Hypatia, 31*(3), 605–619.

Green, S. E. (2007). "We're tired, not sad": Benefits and burdens of mothering a child with a disability. *Social Science & Medicine, 64*(1), 150–163.

Haque, Z., Becares, L., & Treloar, N. (2020). *Over-exposed and under protected. The devastating impact of COVID-19 on black and minority ethnic communities in Great Britain*. Runnymede.

Harry, B. (2008). Collaboration with culturally and linguistically diverse families: Ideal versus reality. *Exceptional Children, 74*(3), 372–388.

Landsman, G. (2005). Mothers and models of disability. *Journal of Medical Humanities, 26*(2–3), 121–139.

Muslim Council of Britain. (2018). Women in mosques development programme (WIMDP). Retrieved December 15, 2020, from https://mcb.org.uk/project/women-in-mosques-development-programme/

Muslim Council of Britain & the University of Cardiff Islam-UK Centre. (2020). #WomenInMosques conversation toolkit. Retrieved January 1, 2021, from https://mcb.org.uk/wp-content/uploads/2020/01/Women_Mosque_Conversation_Toolkit.pdf

Rogers, C. (2011). Mothering and intellectual disability: Partnership rhetoric? *British Journal of Sociology of Education, 32*(4), 563–581.

Runswick-Cole, K., & Ryan, S. (2019). Liminal still? Unmothering disabled children. *Disability & Society, 34*(7–8), 1125–1139.

Women's Budget Group. (2019). Triple whammy: The impact of local government cuts on women. Retrieved December 15, 2020, from https://wbg.org.uk/wp-content/uploads/2019/03/Triple-Whammy-the-impact-of-local-government-cuts-on-women-March-19.pdf

Yosso, T. J. (2005). Whose culture has capital? A critical race theory discussion of community cultural wealth. *Race Ethnicity and Education, 8*(1), 69–91.

YouGov. (2018). Where the public stands on immigration. Retrieved December 15, 2020, from https://yougov.co.uk/topics/politics/articles-reports/2018/04/27/where-public-stands-immigration

Addendum

This book is based on a small-scale qualitative research which provides in-depth narratives of eight British Pakistani Muslim mothers of disabled children, conducted between 2013 and 2017. Throughout the data collection and writing phases of this research, I ensured that my research tools and my broader research decisions were underpinned by a feminist ethical stance. This was challenging in a number of ways and led me to question the extent to which feminist-informed research tools are useful and respectful of minoritised participants (Rizvi, 2017a). For instance, I had initially planned to incorporate maternal feedback within the data analysis stage; however, this was not possible because their everyday mothering responsibilities meant that they had to decline further participation beyond the data collection stage. Nonetheless, mothers did express approval of the interview process itself, and the written transcripts of their interviews that I shared with them were accurate depictions of their interview responses. This illustrates how the materiality of fieldwork and actual interactions with women participants can challenge feminist ideals for engaging in research.

Access

Undertaking this research was a huge responsibility; I constantly questioned every research decision because this research project involved an over researched, heavily scrutinised, silenced, problematized and

S. Rizvi, *Undoing Whiteness in Disability Studies*, https://doi.org/10.1007/978-3-030-79573-3

discriminated community. My research was conducted in Southwest England. I had initially planned to contact both mainstream and special schools with South Asian disabled pupils. However, I eventually decided to pursue an informal approach by focusing on local parent support groups and community centres which I felt would increase the willingness of potential participants to take part in my research. I had initially planned to include British Pakistani and British Bangladeshi mothers in my sample; however, I was unable to establish contact with any Bangladeshi families through local support groups. My participants—mothers and aunts who were primary carers of disabled children—were ultimately contacted through 'Anokha',[1] a minority disabled families' support group whose members were exclusively Pakistani. The mothers, who primarily spoke Punjabi and Urdu, were either first-generation or second-generation immigrants. Since I am fluent in Urdu, I conducted interviews in Urdu unless mothers specified otherwise. Initially, twelve mothers responded with interest to my research introduction letter, which I had sent to all families registered with Anokha. However, once I was 'in the field', I realised that four respondents had only consented to take part due to their asymmetric power dynamics with Anokha as the gatekeepers and so, I ultimately decided that it would be unethical to include them. Although I did not critically examine this at the time, as further participants emerged through snowballing, gatekeeper bias became increasingly apparent. One mother—who will remain unnamed—was rejected as a suitable participant for my research by Anokha because she had a poor relationship with them; I ultimately accessed this mother through snowballing. This particular mother complained that Anokha had delayed her access to formal services by actively hindering signposting of government provisions, and created mistrust about her family within the local British Pakistani community; hence, including this mother in my study increased my sample diversity (McAreavey & Das, 2013). Interestingly, the mothers from the snowballing sample were more culturally, educationally, socially and geographically diverse, and felt separate from Anokha's 'community'. Ultimately, eight mothers took part in this research to share their experiences.

---

[1] Pseudonyms have been used to maintain confidentiality. Anokha is an Urdu word meaning unique.

The Children

The following table provides an overview of the children[2] and their school settings.

Table 1

Mother's Name	Child's Name	SEND Level of Support	School Trajectory
Tahira	Farrukh— (9 years)	Statement	Cavendish Special School- Chester Secondary Special School
Maria	Aamir— (14 years)	Statement	Camphill Mainstream Nursery-Brindley Special Primary and Secondary School
Shehnaz	Amna— (9 years)	Statement	Mainstream Primary school till Y3- Chester Primary Special School
Shehnaz	Tariq—(6 years)	Statement	Cavendish Special School
Saira	Faraz— (19 years)	Statement	Mainstream Primary-Southall Primary Special School- New bridge Academy Secondary Special School
Saira	Farha— (16 years)	School Support	Canterbury Girls Trust Mainstream Secondary School
Saira	Zara—(6 years)	School Support	Darwin Primary Mainstream School
Parveen	Sehr— (deceased) (11 years)	Statement	Cavendish Special School
Parveen	Amber— (6 Years)	Statement	Hilton Mainstream Primary School
Kiran	Ahmed— (15 years)	School- Statement	Rainbow Specialist Nursery- Chester Special School-Primary and Secondary
Alina	Imran— (8 years)	Statement	Dew Mainstream Nursery- Norton Special School
Maham	Daniel— (7 years)	Statement	Dew Mainstream Nursery- Robin Academy Special Primary school

Source: Rizvi (2018, p. 63)

Data Collection

The data collection phase lasted six months, incorporating over 50 hours of interviews. Each participant was interviewed three times, which permitted sufficient time to build a mutual rapport and affinity with my participants; after the data collection phase had ended, I continued to maintain

---

[2] Pseudonyms have been used for children and their schools to protect anonymity.

contact with those mothers who wanted to stay in touch. The use of research diaries, written interview transcripts and field notes also aided me in the writing of this research. The following table outlines how data was collected in this research.

Table 2

Research Phase	Action
Phase 1: Introduction	I met with all the stakeholders (supervisors, gatekeepers, parent participants), and discussed the background and rationale for research, and the potential benefits/costs of taking part in this research. I then asked permission for a voluntary consent from all mother-participants to take part in this research, ensuring they understood that they could leave whenever they wished. (Incidentally, I continued to seek the mothers' consent about participating in the study at each data collection stage.) I used snowballing after I realised that certain participants provided by the gatekeepers would pose ethical concerns around consent
Phase 1: Research Design	I designed my interview questions based on existing understanding and knowledge from current literature on minoritised disabled families, and from feedback and topics/issues mentioned by maternal interviews. I also requested and received permission from mothers to record and use these interviews for analysis and publication later
Phase 2: Data Collection and Revisiting Field	Phase One explored maternal experiences of supporting their children's disability through unstructured interviews. Phase Two examined maternal understanding of their roles within home-school relationships through semi-structured interviews. Phase Three used semi-structured interviews to consider maternal placement preferences, and used vignettes to explore how broader influences like religion, culture, gender and immigration trajectory affect their support. Vignettes were not initially part of my data collection tools, however, after some deliberation, I chose to utilise vignettes, finding them to be an appropriate and respectful research tool to understanding how these wider social realities affected my participants' day-to-day support. Progression from one phase to the next was only done when all participants had been interviewed in each phase
Phase 3: Data Analysis	I shared copies of each mother's interview transcripts with them confidentially, and offered them the opportunity to share individually what they thought about their interviews and how close their transcript was to their own understanding of their interview responses. All mothers were provided with transcripts of each interview before the next interview phase began. In this way, I was able to ask them for any feedback on the transcriptions of previous interviews. I also asked them to participate in the analysis stage of my study; however, all mothers declined to participate any further due to their already hectic schedules. Finally, I carried out a thematic analysis to develop codes and themes

Source: Rizvi (2017b, p. 83)

ETHICAL CONSIDERATION

Engaging with each mother gave me an awareness of their innermost concerns and discourses, family disagreements and their grievances against their children's schools. This huge ethical responsibility only increased after I left the 'field' and I had to report my findings using 'private voices in public discourses', including experiences relevant to my research but inevitably excluding others (Ribbens & Edwards, 1998). I was expected to write my findings in a way that fulfilled institutional requirements whilst not compromising on my feminist stance. Such concerns made this research a very reflective experience, as I endeavoured to present my participants' accounts as ethically and accurately as possible. For further reading, I have also written about this research process in two separate methodology papers (Rizvi, 2017a, 2019).

In my research experience, I have always practiced honesty, authenticity and candour about why I am present in my participants' lives, and the impermanence of my presence, which I have found has managed their expectations about my role. Therefore, I clearly explained my research goals and its practical applicability to my participants' everyday lives. I consulted the BERA guidelines and my university's ethics policy before commencing the research process. There were some significant ethical issues that I wanted to keep at the front of my mind, such as obtaining my participants' consent personally and making sure that they knew they could withdraw their consent at any stage, caring for the mothers' physical safety and emotional wellbeing, and delivering total participant confidentiality. I subsequently ensured that I requested and acquired consent from each of my participants on a one-to-one basis, although the manner of their consent differed. Researchers often assume that all participants are comfortable with English or another language and will have little problem with traditional forms of giving research consent. However, some of my participants expressed discomfort with signing a 'written document'; consequently, I asked for and obtained their verbal consent on audio, whilst assuring them that they could leave the research at any stage, and would not be pressured into consenting against their will by either Anokha or me.

I made certain provisions during my research process in general, and during the data collection phase in particular, to make sure that I did not knowingly put the mothers in my research at physical risk or distress. Therefore, I often held participant interviews at times which suited the mothers schedules and at locations which safeguarded their privacy. For instance, Tahira lived with her in-laws, who closely monitored and scrutinised her behaviour and actions; as a result, we agreed to hold our interviews when her in-laws were away from home. This meant that I had to be ready to visit her home at short notice to interview her whenever she texted me. I also prioritised my participants confidentiality by assigning them a pseudonym, and only used the mothers' real names and contact details when I was contacting them to schedule interviews or obtaining their consent. Whilst Anokha were acquainted with the names and contact details of the mothers they had suggested to me—and which unquestionably affected the anonymity of those mothers—by giving each of my participants a pseudonym during my data analysis phase and in my final report and by anonymising their children's names reduced the negative impact somewhat. Moreover, the mothers were given repeated assurances that no one including Anokha, education, health or social care professionals, or other members of their local community would be able to read or hear their interviews. They were also assured that any information which they did not wish to share would not be disclosed. Finally, I also designed my interview questions and talking points in a way that did not scrutinise my participants' parenting skills or question their decisions, or in any way cause intentional or unnecessary pain or anguish.

For my own safety, I restricted all information about the names, times and places of my interviews to a few colleagues, which I secured in a password-protected Google Cloud drive. I also deleted all digital audio files of my participant interviews as soon as I completed my research report.

Maintaining ethical considerations at the forefront of my research project was challenging, and very much a feature that ran throughout my research process. However, I always tried my utmost to conduct every aspect of my research in a professional, responsible and ethical manner across my research journey.

References

McAreavey, R., & Das, C. (2013). A delicate balancing act: Negotiating with gatekeepers for ethical research when researching minority communities. *International Journal of Qualitative Methods, 12*(1), 113–131.

Ribbens, J., & Edwards, R. (1998). Living on the edges: Public knowledge, private lives, and personal experience. In J. Ribbens & E. Rosalind (Eds.), *Feminist dilemmas in qualitative research: Public knowledge and private lives. Social research methods online* (pp. 2–20). Sage.

Rizvi, S. (2017a). Treading on eggshells: 'Doing' feminism in educational research. *International Journal of Research & Method in Education, 42*(1), 46–58.

Rizvi, S. (2017b). *Fighting your corner: An in-depth study of how British-Pakistani mothers support their child with SEND.* Doctoral dissertation, Graduate School of Education, University of Bristol. https://bris.on.worldcat.org/oclc/1052802226

Rizvi, S. (2018). There's never going to be a perfect school that ticks every box: Minority perspectives of inclusion and placement preferences. *Journal of Research in Special Educational Needs, 18*, 59–69.

Rizvi, S. (2019). Using fiction to reveal truth: Challenges of using vignettes to understand participant experiences within qualitative research. *Forum: Qualitative Sozialforschung/Forum: Qualitative Social Research, 20*(1), Art. 10. https://doi.org/10.17169/fqs-20.1.3101

Index[1]

[1] Note: Page numbers followed by 'n' refer to notes.